PUBLISHER COMMENTARY

We print NASA's handbooks and standards for the convenience of those that use them on a daily basis. We print all of these a full 8 ½ by 11 with large text so they are easy to read. Yes, color books are expensive to print so unless the information relies on the use of color for proper interpretation or understanding, we print most books in black and white to keep the cost down. All these documents are available for download for free from NASA, however printing them all over a network printer would take days.

Why buy a book you can download free? We print this so you don't have to.

All these books are available for free download from the government web site. Some are available only in electronic media. Some online docs are missing pages or barely legible.

We at 4th Watch Publishing are former government employees, so we know how government employees actually use the standards. When a new standard is released, an engineer prints it out, punches holes and puts it in a 3-ring binder. While this is not a big deal for a 5 or 10-page document, many NIST documents are over 100 pages and printing a large document is a time-consuming effort. So, an engineer that's paid $75 an hour is spending hours simply printing out the tools needed to do the job. That's time that could be better spent doing engineering. We publish these documents so engineers can focus on what they were hired to do – engineering. It's much more cost-effective to just order the latest version from Amazon.com

If there is a standard you would like published, let us know. Our web site is www.usgovpub.com

www.usgovpub.com

List of Other NASA Publications Available on Amazon.com:

NASA-STD-5001B	Structural Design and Test Factors of Safety for Spaceflight Hardware
NASA-STD-5006A	General Welding Requirements for Aerospace Materials
NASA-STD-5008B	Protective Coating of Carbon Steel, Stainless Steel, and Aluminum on Launch Structures, Facilities, and Ground Support Equipment
NASA-STD-5009A	Nondestructive Evaluation Requirements for Fracture-Critical Metallic Components
NASA-STD-5012B	Strength and Life Assessment Requirements for Liquid-Fueled Space Propulsion System Engines
NASA-STD-5019A	Fracture Control Requirements for Spaceflight Hardware
NASA-STD-5005D	Standard for The Design and Fabrication of Ground Support Equipment
NASA-HDBK-8739.21	Workmanship Manual for Electrostatic Discharge Control
NASA-HDBK 8739.23A	NASA Complex Electronics Handbook for Assurance Professionals (Color)
NASA-HDBK-8719.14	Handbook for Limiting Orbital Debris (Color)
NASA-HDBK-8709.22	Safety and Mission Assurance Acronyms, Abbreviations, and Definitions
NASA-HDBK-7009	NASA Handbook for Models and Simulations: An Implementation Guide For NASA-STD-7009 (Color)
NASA-HDBK-8739.19-2	Measuring and Test Equipment Specifications NASA Measurement Quality Assurance Handbook – Annex 2
NASA-HDBK-8739.19-3	Measurement Uncertainty Analysis Principles and Methods NASA Measurement Quality Assurance Handbook – Annex 3
NASA-HDBK-8739.19-4	Estimation and Evaluation of Measurement Decision Risk NASA Measurement Quality Assurance Handbook – Annex 4
NASA RCM	Reliability-Centered Maintenance Guide for Facilities and Collateral Equipment

www.usgovpub.com

Approved: 2010-07-13
Baseline

NASA HANDBOOK

Measuring and Test Equipment Specifications

NASA Measurement Quality Assurance Handbook – ANNEX 2

Measurement System Identification: **Metric**

July 2010

National Aeronautics and Space Administration
Washington DC 20546

This page intentionally left blank.

DOCUMENT HISTORY LOG

Status	Document Revision	Approval Date	Description
Baseline		2010-07-13	Initial Release *(JWL4)*
	Revalidated	2018-03-01	Baseline revalidated.

This document is subject to reviews per Office of Management and Budget Circular A-119, Federal Participation in the Development and Use of Voluntary Standards (02/10/1998) and NPD 8070.6, Technical Standards (Paragraph 1.k).

This page intentionally left blank.

FOREWORD

This publication provides principles and methods for the analysis and management of the uncertainty that exists in measurements. The principles and methods described herein support Agency objectives for assuring measurement accuracy and are applicable for use in all NASA functions for which measurements and decisions based on measurements are involved.

This Annex to NASA-HDBK 8739.19 is approved for use by NASA Headquarters and NASA Centers, including Component Facilities. This document may be referenced on contracts as a guidance or training publication.

Comments and questions concerning the contents of this publication should be referred to the National Aeronautics and Space Administration, Director, Safety and Assurance Requirements Division, Office of Safety and Mission Assurance, Washington, DC 20546.

Requests for information, corrections, or additions to this NASA HDBK shall be submitted via "Feedback" in the NASA Technical Standards System at http://standards.nasa.gov or to National Aeronautics and Space Administration, Director, Safety and Assurance Requirements Division, Office of Safety and Mission Assurance, Washington, DC 20546.

Bryan O'Connor   13 JULY 2010
Chief, Safety and Mission Assurance   Approval Date

This page intentionally left blank.

TABLE OF CONTENTS

LIST OF FIGURES

LIST OF TABLES

ACKNOWLEDGEMENTS

The principal author of this publication is Ms. Suzanne Castrup of Integrated Sciences Group (ISG), Inc. of Bakersfield, CA under NASA Contract.

Additional contributions and critical review were provided by:

Scott M. Mimbs
NASA Kennedy Space Center

James Wachter,
MEI Co., NASA Kennedy Space Center

Charles P. Wright
The Aerospace Corporation

Dave Deaver
Fluke Corporation

Mihaela Fulop
SGT, NASA Glenn Research Center, OH

Roger Burton
Honeywell International, Inc.

Steve Lewis
Lockheed Martin Mission Systems & Sensors

Dr. James Salsbury
Mitutoyo America Corporation

This Publication would not exist without the recognition of a priority need by the NASA Metrology and Calibration Working Group and funding support by NASA/HQ/OSMA and NASA/Kennedy Space Center.

EXECUTIVE SUMMARY

Manufacturer specifications are an important element of cost and quality control for testing, calibration and other measurement processes. They are used in the selection of measuring and test equipment (MTE) and in the establishment of suitable equipment substitutions for a given measurement application. In addition, manufacturer specifications are used to estimate bias uncertainties and establish tolerance limits for MTE attributes and parameters.

MTE attributes and parameters are periodically calibrated to determine if they are in conformance with manufacturer specified tolerance limits. The elapsed-time or interval between calibrations is often based on in-tolerance or out-of-tolerance data acquired from periodic calibrations. Therefore, it is important that manufacturer specifications provide sufficient information about the MTE performance attributes and parameters and that the specifications are properly interpreted and applied in the establishment of test tolerance limits or the estimation of measurement uncertainty.

Manufacturers of complex instruments often include numerous time and range dependent specifications. And, since specification documents are also a means for manufacturers to market their products, they often contain additional information about features and operating condition limits that must be carefully reviewed and understood.

Unfortunately, there are no universal guidelines or standards regarding the content and format of MTE specification documents. This is evident from the inconsistency in the manufacturer specifications for similar equipment. Some manufacturers provide specifications for individual performance parameters, while others provide a single specification for overall accuracy. Inconsistencies in the development of MTE specifications, and in the terms and units used to convey them, create obstacles to their proper interpretation.

This document provides an in-depth discussion about developing, verifying and reporting MTE specifications. Recommended practices are presented and illustrative examples are given for interpreting and applying manufacturer specifications to assess MTE performance and reliability.

Reader's Guide

The methods and practices presented herein are primarily intended for technical personnel responsible for the development, selection, application and support of MTE. However, this document may also be useful for project managers, scientists and engineers that rely on measurement data for their analysis, evaluation and decision making processes.

Chapters 1 through 3 are intended for all personnel. Chapter 1 presents the purpose and scope of this document, along with some introductory material regarding MTE specification documents, their interpretation and application. Chapter 2 presents a brief overview of the various types of MTE addressed in this document. Chapter 3 discusses the static and dynamic performance characteristics common to a wide variety of MTE and the environmental operating conditions that can affect MTE performance.

Chapters 4 through 7 are intended for personnel responsible for the development and manufacture of MTE. Chapters 4 and 5 discuss the analysis and testing methods used to establish MTE performance characteristics and develop specification tolerance limits. Chapter 6 discusses acceptance testing, production monitoring and engineering analysis methods used to verify and modify MTE specifications. Chapter 7 presents recommended practices for reporting MTE specifications, including some practical content and format guidelines.

Chapters 8 through 12 are intended for personnel responsible for MTE selection, application and support. Chapter 8 discusses the various sources for obtaining MTE specifications and related performance information. Chapter 9 presents guidelines for interpreting MTE specifications and provides illustrative examples for combining specifications. Chapter 10 discusses the role of calibration in the validation of MTE performance capabilities and assuring measurement quality and reliability.

Chapter 11 provides illustrative examples of how MTE specifications are used to estimate parameter bias uncertainties, compute test tolerance limits, determine in-tolerance probability, and establish calibration intervals. Chapter 12 addresses issues concerning firmware and software-based specifications.

Chapter 13 is intended for all personnel. Chapter 13 discusses the role of MTE specifications in assuring measurement quality and provides recommended practices for defining measurement requirements, designing measurement systems and selecting MTE.

ACRONYMS AND INITIALISMS

ADC	Analog to Digital Converter
AICHE	American Institute of Chemical Engineers
ALT	Accelerated Life Testing
AOP	Average-over-Period
ASME	American Society of Mechanical Engineers
ASTM	American Society for Testing and Materials
BOP	Beginning-of-Period
BFSL	Best-fit Straight Line
CMR	Common-Mode Rejection
CMRR	Common-Mode Rejection Ratio
CMV	Common-Mode Voltage
COTS	Commercial Off-the-Shelf
CTE	Coefficient of Thermal Expansion
CxP	Constellation Program
DAC	Digital to Analog Converter
DMM	Digital Multimeter
DOD	Department of Defense
EOP	End-of-Period
FAA	Federal Aviation Administration
FS	Full Scale
FSI	Full Scale Input
FSO	Full Scale Output
FSR	Full Scale Range
GIDEP	Government Industry Data Exchange Program
HALT	Highly Accelerated Life Test
HASS	Highly Accelerated Stress Screening
HVAC	Heating, Ventilating and Air Conditioning
IEC	International Electrotechnical Commission
ISA	International Society for Instrumentation, Systems and Automation
ISO	International Organization for Standardization
LSB	Least Significant bit
LSD	Least Significant Digit
LVDT	Linear Variable Differential Transformer
MTE	Measuring and Test Equipment
MTBF	Mean Time Between Failures
NASA	National Aeronautics and Space Administration
NIST	National Institute of Standards and Technology (U.S.)
NMRR	Normal-Mode Rejection Ratio
NMV	Normal-Mode Voltage
NPL	National Physical Laboratory (U.K.)
NSTS	NASA Space Transportation Systems
OOT	Out of Tolerance
PDF	Probability Density Function
	Portable Document Format (Adobe Acrobat Files)
ppb	Parts per billion
ppm	Parts per million
PTB	Physikalisch-Technische Bundesanstalt (German Metrology Institute)

RDG	Reading
RF	Radio Frequency
RH	Relative Humidity
RO	Rated Output
RSS	(1) Root-sum-square method of combining values.
	(2) Residual sum of squares in regression analysis.
RTI	Referred to Input
RTO	Referred to Output
RVDT	Rotational Variable Differential Transformer
SI	International System of Units
SMA	Scale Manufacturers Association
TRS	Transverse Rupture Strength
UUT	Unit Under Test
VAC	Alternating Current Volts
VDC	Volts Direct Current
VIM	International Vocabulary of Basic and General Terms in Metrology

CHAPTER 1: INTRODUCTION

Ideally, manufacturer specifications should provide performance characteristics that can be used to evaluate the suitability of MTE for a given application. However, understanding specifications and using them to compare equipment from different manufacturers or vendors can be a perplexing task. This primarily results from inconsistent terminology, units and methods used to develop and report equipment specifications.

Manufacturer specifications are used to compute test uncertainty ratios and estimate bias uncertainties essential for measurement uncertainty analysis and decision risk analysis. In addition, MTE are periodically calibrated to determine if they are performing within manufacturer specified tolerance limits. Therefore, it is important that MTE specifications are properly reported, interpreted and applied.

1.1   Purpose

This document is intended to provide recommended practices for developing, reporting, obtaining, interpreting, validating and applying MTE specifications. Detailed examples are included to educate, demonstrate, promote and reinforce best practices.

1.2   Scope

This document presents technical information and best practices relevant to MTE development, selection, acceptance testing, calibration and end-use.

This document is targeted to various MTE developers and users, including:

- Scientists and engineers responsible for the development of leading-edge measurement technology.

- System designers that configure various MTE components to meet desired measurement requirements.

- Personnel responsible for purchasing or selecting MTE based on published specifications.

- Metrology and calibration personnel responsible for the validation and maintenance of MTE performance.

1.3   Background

For the most part, specifications are intended to convey tolerance or confidence limits that are expected to bound MTE parameters or attributes. For example, these limits may correspond to temperature, shock and vibration parameters that affect the sensitivity and/or zero offset of a sensing device.

Manufacturer specifications are used in the purchase or selection of substitute MTE for a given measurement application, the estimation of bias uncertainties and the establishment of tolerance limits for calibration and testing. Therefore, MTE users must be proficient at identifying applicable specifications and in interpreting and applying them.

MTE specifications should provide adequate details about performance characteristics for a representative group of devices or items (e.g., manufacturer/model). This information should be reported in a logical format, using consistent terms, abbreviations and units that clearly convey pertinent parameters or attributes. Given the complexity of present day measuring equipment and instruments there is a need for standardized specification formats.

Technical organizations, such as ISA and SMA, have published documents that adopt standardized instrumentation terms and definitions. ISA has also developed standards that provide uniform requirements for specifying design and performance characteristics for selected electronic transducers.[1] Similarly, SMA has developed a standard for the data and specifications that must be provided by manufacturers of load cells.

Despite these few exceptions for electronic transducers, the vast majority of specification documents fall short of providing crucial information needed to evaluate MTE for a given measurement application. It is common for manufacturers to omit information about the underlying probability distributions for the MTE performance parameters. In addition, manufacturers don't often report the corresponding confidence level for the specified MTE parameter tolerance limits.

The lack of universal guidelines or standards regarding the content and format of MTE specification documents is evident from the inconsistency in the manufacturer specifications for similar equipment.[2] These inconsistencies create obstacles to the proper interpretation and application of MTE specifications.

Although MTE specifications are an important element of measurement quality assurance, only a handful of articles and papers have been written about the difficulties encountered when interpreting MTE specifications. The measurement science community as a whole is only now beginning to formally address the issues regarding their development, interpretation and application.

Consequently, there is a need for comprehensive documentary guidance and recommended practices for developing, reporting, interpreting, validating and applying MTE specifications. This *Measurement Quality Assurance Handbook* Annex covers these topics and provides references to relevant external standards, recommended practices and guidance documents.

1.4   Application

The recommended practices, procedures and illustrative examples contained in this document provide a comprehensive resource for MTE developers, users and calibration personnel. This document should be used as a companion to the following NASA *Measurement Quality Assurance Handbook* Annexes:

- ▸ Annex 1 - *Measurement Quality Assurance End-to-End.*
- ▸ Annex 3 - *Measurement Uncertainty Analysis Principles and Methods.*
- ▸ Annex 4 - *Estimation and Evaluation of Measurement Decision Risk.*
- ▸ Annex 5 - *Establishment and Adjustment of Calibration Intervals.*

---

[1] A list of ISA transducer standards is given in Appendix B.

[2] Similar equipment constitutes MTE from different manufacturers that can be substituted or interchanged without degradation of measurement capability and quality.

CHAPTER 2: MEASURING AND TEST EQUIPMENT

Before we delve into developing, obtaining and interpreting specifications, it is important to clarify what constitutes MTE. In the fields of measurement science and metrology, MTE include artifacts, instruments, sensors and transducers, signal conditioners, data acquisition units, data processors, output displays, cables and connectors.

2.1   Artifacts

Artifacts constitute devices such as mass standards, standard resistors, pure and certified reference materials, gage blocks, etc. Accordingly, artifacts have stated outputs or nominal values and associated specifications.

2.2   Instruments

Instruments constitute equipment or devices that are used to measure and/or provide a specified output. They include, but are not limited to, oscilloscopes, wave and spectrum analyzers, Josephson junctions, frequency counters, multimeters, signal generators, simulators and calibrators, inclinometers, graduated cylinders and pipettes, spectrometers and chromatographs, micrometers and calipers, coordinate measuring machines, balances and scales. Accordingly, instruments can consist of various components with associated specifications and tolerance limits.

2.3   Sensors and Transducers

Sensors constitute equipment or devices that detect or respond to a physical input such as pressure, acceleration, temperature or sound. The terms sensor and transducer are often used interchangeably. Transducers more generally refer to devices that convert one form of energy to another. However, while all sensors are transducers, all transducers are not sensors. For example, actuators that convert an electrical signal to a physical output are also considered to be transducers. For the purposes of this document, discussion will be limited to sensors and transducers that convert a physical input to an electrical output.

> **Note:** Transmitters constitute sensors coupled with internal signal conditioning and/or data processing components, as well as an output display.

Some sensors and transducers convert the physical input directly to an electrical output, while others require an external excitation voltage or current. Sensors and transducers encompass a wide array of operating principles (i.e., optical, chemical, electrical) and materials of construction. Therefore, their performance characteristics and associated specifications can cover a broad spectrum of detail and complexity. A selected list of sensors and transducers is shown in Table 2-1.

Table 2-1. Sensors and Transducers

Input	Sensor/Transducer	Output	Excitation
Temperature	Thermocouple	Voltage	
	RTD	Resistance	Current
	Thermistor	Resistance	Current, Voltage
Pressure and Sound	Strain Gauge	Resistance	Voltage
	Piezoelectric	Voltage	
Force and Torque	Strain Gauge	Voltage	Voltage

Input	Sensor/Transducer	Output	Excitation
	Piezoelectric	Voltage	
Acceleration/Vibration	Strain Gage	Voltage	Voltage
	Piezoelectric	Charge	
	Variable Capacitance	Voltage	Voltage
Position/Displacement	LVDT and RVDT	AC Voltage	Voltage
	Potentiometer	Voltage	Voltage
Light Intensity	Photodiode	Current	
Flow Rate	Coriolis	Frequency	
	Vortex Shedding	Pulse/Frequency	Voltage
	Turbine	Pulse/Frequency	
pH	Electrode	Voltage	

2.4   Signal Conditioners

Signal conditioners constitute devices or equipment employed to modify the characteristic of a signal. Conditioning equipment include attenuators, amplifiers, bridge circuits, filters, analog-to-digital and digital-to-analog converters, excitation voltage or current, reference temperature junctions, voltage to frequency and frequency to voltage converters, multiplexers and linearizers. A representative list of signal conditioning methods and functions is provided in Table 2-2.

Table 2-2. Signal Conditioning Methods

Type	Function
Analog-to-Digital Conversion (ADC)	Quantization of continuous signal.
Amplification	Increase signal level.
Attenuation	Decrease signal level.
Bridge Circuit	Used in the measurement of various electrical quantities, including resistance. The Wheatstone bridge is a commonly used bridge circuit.
Charge Amplification	Change integrated current to voltage.
Cold Junction Compensation	Provide temperature correction for thermocouple connection points.
Digital-to-Analog Conversion (DAC)	Convert discrete signal to continuous signal.
Excitation	Provide voltage or current to non-self generating transducers.
Filter	Provide frequency cutoffs and noise reduction.
Isolation	Block high voltage and current surges.
Linearization	Convert non-linear signal to representative linear output.
Multiplexing	Provide sequential routing of multiple signals.

2.5   Data Acquisition Equipment

Data acquisition (DAQ) equipment constitute devices that gather (acquire) and store measured data or signals. DAQ equipment include computers, data loggers, remote terminal units, high-speed timers, random access memory (RAM) devices and USB flash drives.

2.6   Data Processors

Data processors constitute equipment or methods used to implement necessary calculations. Data processors include totalizers, counters and computers incorporating statistical methods, regression or curve fitting algorithms, interpolation schemes, measurement unit conversion or other computations.

2.7   Output Displays

Output display devices constitute equipment used to visually present processed data. Display devices can be analog or digital in nature. Analog devices include chart recorders, plotters and printers, dials and gages, cathode ray tube (CRT) panels and screens. Digital devices include light-emitting diode (LED) and liquid crystal display (LCD) panels and screens.

2.8   Cables and Connectors

Additional, ancillary MTE include cables and connectors used during the measurement process. electrical wire, fiberoptic and coaxial cables, connectors, splitters, hubs, switches, adapters and couplers.

CHAPTER 3: PERFORMANCE CHARACTERISTICS

Manufacturer specifications describe the MTE performance characteristics, parameters or attributes that are covered under product warranty. Depending on the type of MTE, manufacturers may include both static and dynamic performance characteristics. And, since specification documents are also a means for manufacturers to market their products, they often contain additional information about features, operating condition limits, or other qualifiers that establish warranty terms.

Some manufacturers may provide ample information detailing individual performance specifications, while others may only provide a single specification for overall accuracy. In some instances, specifications can be complicated, including numerous time dependent, range dependent or other characteristics.

3.1 Static Characteristics

Static performance characteristics provide an indication of how an instrument, transducer or signal conditioning device responds to a steady-state input at one particular time. In addition to sensitivity (or gain) and zero offset, other static characteristics include nonlinearity, repeatability, hysteresis, resolution, noise and accuracy.[3]

3.1.1 Sensitivity

Sensitivity is defined as the ratio of the output signal to the corresponding input signal for a specified set of operating conditions. Similarly, gain is the ratio of the amplifier output signal voltage to the input signal voltage. If the amplification ratio is less than unity, then it is called the attenuation.

The sensitivity of a measuring device or instrument depends on the principle of operation and the design. Many devices or instruments are designed to have a linear relationship between input and output signals and thus provide a constant sensitivity over the operating range. As a result, MTE manufacturers often report a nominal or ideal sensitivity with a stated error or accuracy.

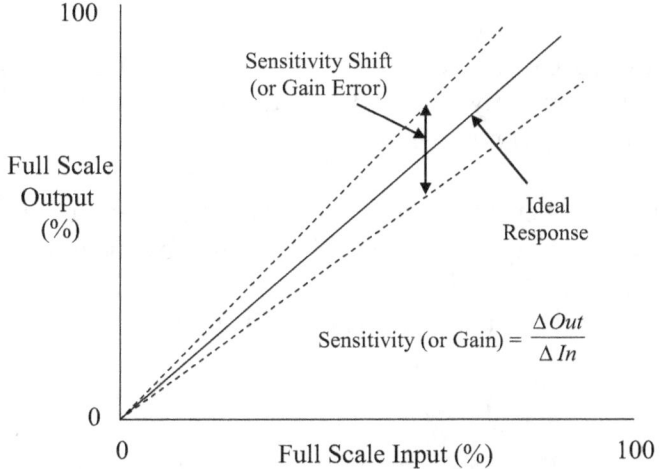

Figure 3-1 Sensitivity and Sensitivity Error

---

[3] Accuracy is often reported as a combined specification that accounts for MTE nonlinearity, hysteresis, and repeatability performance specifications.

As shown in Figure 3-1, the actual output response may be linear but may differ from the specified nominal or ideal sensitivity. The difference between the actual and ideal output response is the sensitivity shift or error. A better estimation of the actual sensitivity of the device can be determined by calibration. The associated errors due to nonlinearity, hysteresis and repeatability cannot be eliminated, but the uncertainty due to these errors can be quantified through proper calibration.

3.1.2   Zero Offset

Zero offset occurs if the device exhibits a non-zero output for a zero input. Zero offset is assumed constant at any input level and, therefore, contributes a fixed error throughout the measurement range, as shown in Figure 3-2. Although zero offset error may be reduced by adjustment, there is no way to completely eliminate it because there is no way to know the true value of the offset. Offset error is typically reported as a percent of full scale or in terms of fundamental units such as volts or millivolts.

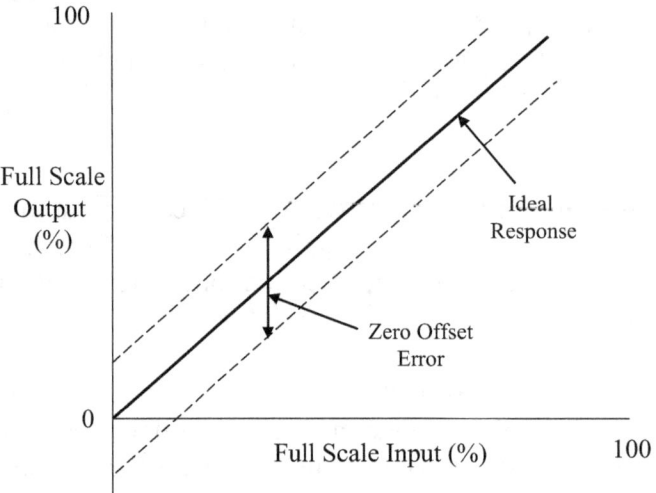

Figure 3-2 Offset Error

3.1.3   Nonlinearity

Nonlinearity is another measure of the deviation of the actual response of the device from an ideal linear relationship. Nonlinearity error exists if the actual sensitivity is not constant over the range of the device, as shown in Figure 3-3.

At any given input, nonlinearity error is fixed, but varies with magnitude and sign over a range of inputs. Nonlinearity (or linearity) error is usually defined by the amount that the output differs from ideal behavior over the full range of the device. Therefore, nonlinearity is often stated as a percentage of the full scale output of the device.

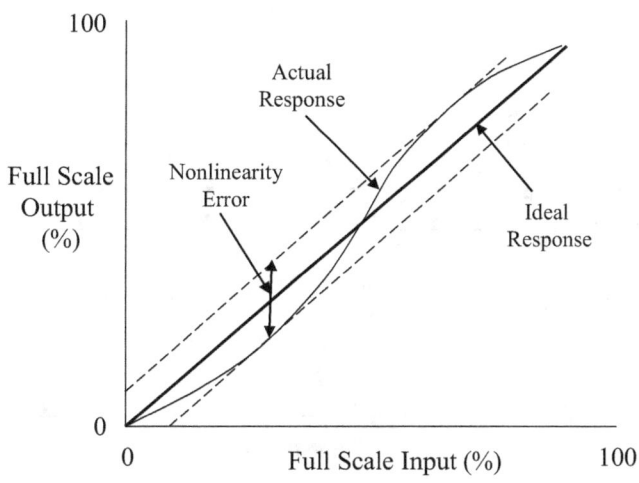

Figure 3-3 Nonlinearity Error

3.1.4 Hysteresis

Hysteresis indicates that the output of the device is dependent upon the direction and magnitude by which the input is changed. At any input value, hysteresis can be expressed as the difference between the ascending and descending outputs, as shown in Figure 3-4. Hysteresis error is fixed at any given input, but can vary with magnitude and sign over a range of inputs. This error is often reported as a percent of full scale.

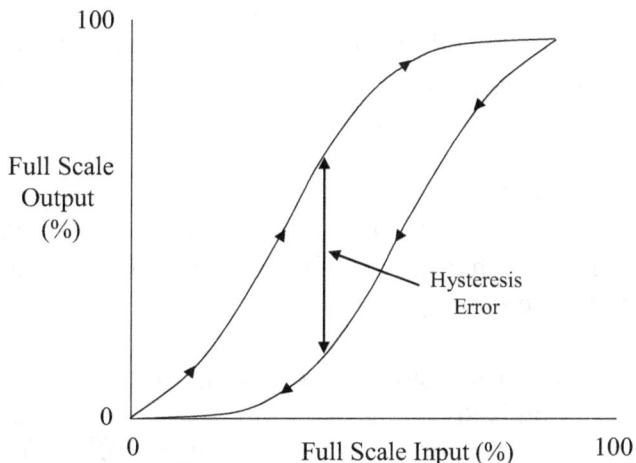

Figure 3-4 Hysteresis Error

3.1.5 Noise

The random error intrinsic to a device or instrument that causes the output to vary from observation to observation for a constant input, as shown in Figure 3-5. In some MTE specifications, non-repeatability and noise are synonymous and are considered to be a short-term stability specification.

Given its random nature, noise varies with magnitude and sign over a range of inputs. Although noise can be reduced by signal conditioning, such as filtering, it cannot be eliminated completely. Noise is typically specified as a percentage of the full scale output.

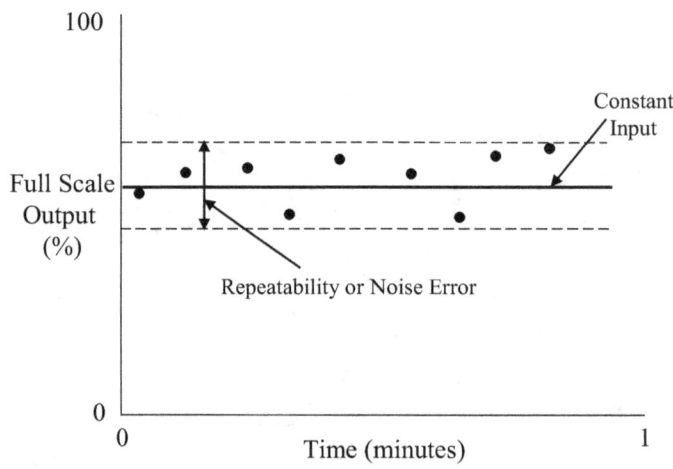

Figure 3-5 Repeatability or Noise Error

3.1.6 Resolution

Resolution is a characteristic that occurs when a measuring device or instrument produces a series of discrete output steps in response to a continuously varying input. Examples include wire-wound potentiometers, turbine flowmeters, analog-to-digital converters (ADC's) and digital displays.

For wire-wound potentiometers, a change in output is only produced when the wiper moves from one wire in the coil to the next adjacent wire. An ADC converts a continuous input signal to a series of quantized or discrete output values. For example, the resolution resulting from the quantization of a 0 to 5 V analog signal using a 12-bit ADC is $5 \text{ V}/(2^{12})$ or 1.2 mV. The quantization error limits are half the resolution or ± 0.6 mV.

3.2 Dynamic Characteristics

It is important that MTE are able to accurately measure time-varying inputs. How a measuring device or instrument responds to a change in input is determined by its design (e.g., materials, parts and components) and its mode of operation.

Dynamic performance characteristics provide an indication of how an instrument, transducer or signal conditioning device responds to changes in input over time. Dynamic characteristics include warm-up time, response time, time constant, settling time, zero drift, sensitivity drift, stability, upper and lower cutoff frequencies, bandwidth, resonant frequency, frequency response, damping and phase shift.

3.2.1 Response Time

Because of mechanical, electrical, thermal or other inertial constraints, the sensing element of a measuring device or instrument cannot respond instantly to changes in input conditions. Response time characterizes the time it takes for the output of a device or instrument to reach a specified percentage of its final value when a step change[4] in the input is applied. The shape of the response time curve will vary for first-order, second-order or higher-order devices or systems. The response time for a first-order device or system is shown in Figure 3-6 for illustration.

---

[4] A step change occurs when the input is increased or decreased very quickly.

9

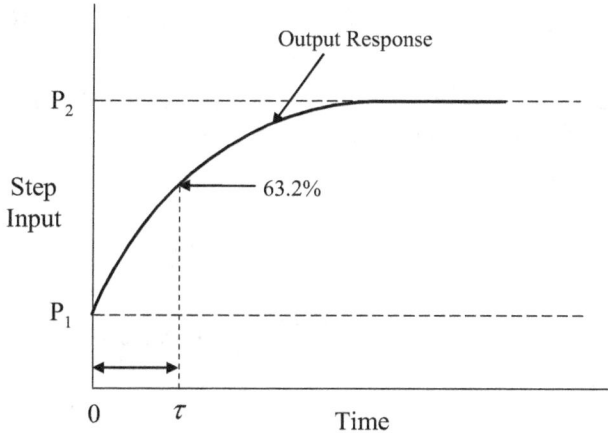

Figure 3-6 Response Time for First-order Device

As shown in Figure 3-6, the time constant, τ, is the specific time required for the output to reach 63.2% of its final value for a step change in input. Similarly, the rise time is the time required for the output to change from 10% to 90% of the final value for a step change in the input.

The rate at which the sensor output finally reaches its final value depends on the sensor design, installation effects, and the magnitude of the input change. For example, in RC circuits, the time constant, in seconds, is $\tau = R \cdot C$, where R is the resistance in ohms and C is the capacitance in farads.

Response time error can occur when the sampling time is insufficient for the sensing device to respond to the input change, as shown in Figure 3-7.

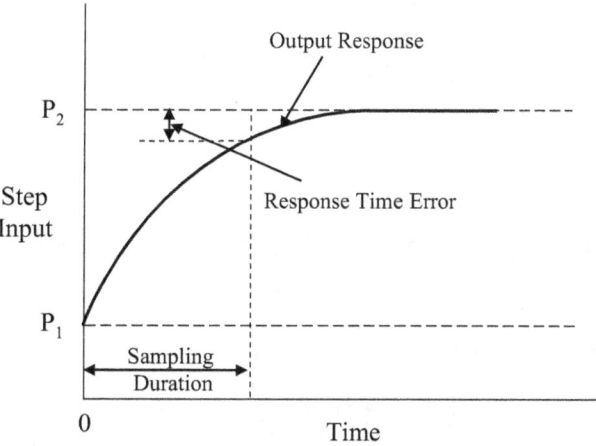

Figure 3-7 Response Time Error for First-order Device

3.2.2 Frequency Response

Amplitude-frequency response, commonly referred to as frequency response, characterizes the change in the ratio of amplitudes and phase differences of the output and input signals. Frequency response is expressed in decibels (dB) for a range of frequencies. Ideally, the frequency response curve should be flat over the frequency range of interest. In reality, some distortion or ripple will occur, as shown in Figure 3-8.

10

Amplitude flatness or ripple specifications establish how constant the output remains over the desired frequency range. Flatness and ripple are usually expressed as ± dB. Stopband is the part of the frequency spectrum in which the signal is subjected to a specified amplitude loss (i.e., attenuated) by a filter. Bandwidth or passband is the range of frequencies in which the response is within 3 dB of the maximum response.

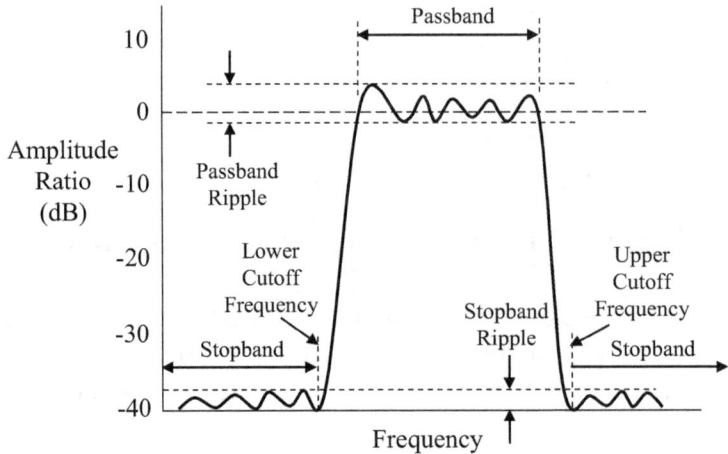

Figure 3-8 Frequency Response Curve

Phase shift characterizes the time lag or delay between the input and output signals, as shown in Figure 3-9. The time delay is a function of the frequency of the input signal and the damping of the measuring device. For resonant systems with sinusoidal input and frequency, f, the time delay is a phase angle, θ, given by

$$\tan\theta = \frac{2\zeta\dfrac{f}{f_n}}{1-\left(\dfrac{f}{f_n}\right)^2} \tag{3-1}$$

where f_n is the natural frequency, ζ is the damping ratio and $\theta/2\pi$ is the number of seconds the output lags behind the input.

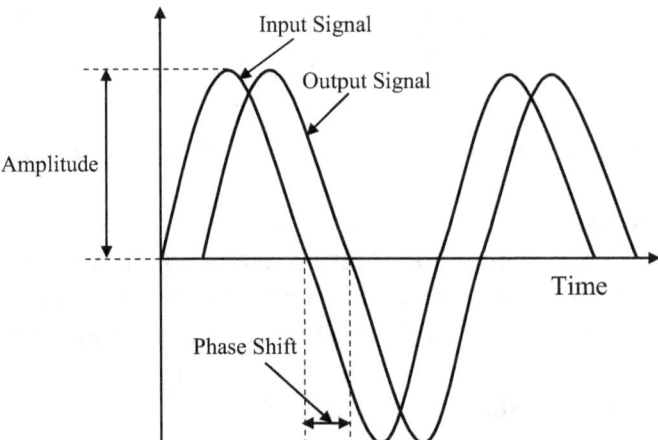

Figure 3-9 Phase Shift

11

3.2.3 Damping Ratio

Damping is a characteristic that defines how energy from a rapid change in input is dissipated within a device or instrument. A device is under-damped if the output exceeds or overshoots the final output value. A critically damped device does not exceed the output final value and a minimum time is required to reach the final value. An over-damped device does not overshoot the final output value, but the response time and the upper limit of the response frequency are adversely impacted. Manufacturers may add damping to a device by electrical or mechanical means to eliminate overshoot and oscillation.

As shown in Figure 3-10, the damping ratio is a measure of how fast oscillations in a system decay over time. If the damping ratio is low (under-damped), the output response can oscillate for a long time. Conversely, if the damping ratio is large (over-damped), the output response may not oscillate at all.

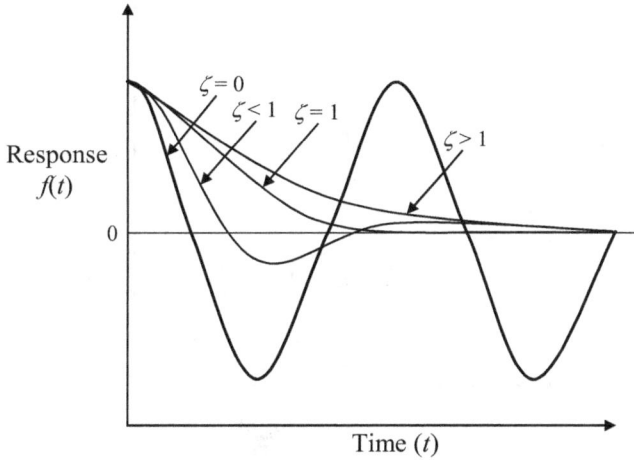

Figure 3-10 Damping Ratio

3.2.4 Stability

Stability specifications provide an indication of the ability of the MTE to maintain its performance characteristics over an extended period of time. Changes in sensitivity and/or zero offset over time, can be a major problem that can affect MTE performance during use.

Consequently, manufacturer specifications usually include a time period during which the MTE parameters can be expected to perform within stated tolerance limits. The time period is provided to account for the drift rate inherent to the device or instrument. In some cases, the manufacturer may specify performance tolerance limits for several time calibration periods (i.e., 30, 90, 180 and 360 days).

As previously discussed, the sensitivity and/or zero offset of a device or instrument can also change or drift due to environmental operating conditions. The influence of time-dependent and environmental influences may be interrelated, in which case they can be difficult to specify separately.

3.3   Other Characteristics

Additional characteristics are often included in MTE specifications to indicate input and output ranges, environmental operating conditions, external power requirements, weight, dimensions and other physical aspects of the device. These other characteristics include rated output, full scale output, range, span, dynamic input range, threshold, deadband, operating temperature range, operating pressure range, operating humidity range, storage temperature range, thermal compensation, temperature compensation range, vibration sensitivity, excitation voltage or current, weight, length, height, and width.

3.3.1 Span, Deadband and Threshold

Manufacturers design their equipment or instruments to make measurements over a specified range of input values. The span or dynamic input range, shown in Figure 3-11, is the algebraic difference between the lower detection limit and the upper detection limit of the device or instrument. The rated or full scale output defines the maximum and minimum output values that the device can provide.

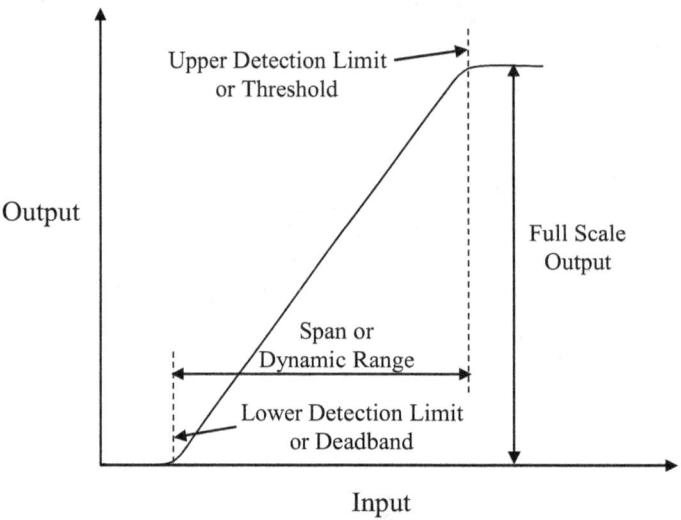

Figure 3-11 Deadband, Threshold, Span and Full Scale Output

The deadband of a device or instrument characterizes the value of the input signal at which a non-zero output occurs. Deadband is usually expressed as a percentage of the span of the device. The threshold or upper detection limit is the maximum output signal of the device, beyond which increases in the input will not result in a change in the output.

3.3.2 Environmental Operating Conditions

The operating environment can have a major effect on MTE performance. Environmental operating conditions include temperature, pressure, humidity, shock, electromagnetic and electrostatic effects, dust and other airborne particulates, and acceleration.

It is common for manufacturers to state the MTE specifications that apply for a given range of climatic, mechanical and electromagnetic conditions. These environmental operating conditions, usually expressed as ± values, also define the conditions in which the MTE can be operated or used without permanently affecting the performance characteristics.

3.3.3 Environmental Effects

MTE manufacturers typically provide baseline or primary specification limits that are applicable for a narrow operating temperature range (e.g., 23°C ± 5°C). This temperature range may be applicable for MTE used in controlled laboratory environments. However, many devices are used in factories, process plants and other environments where the operating temperatures greatly exceed these limits.

In such cases, manufacturers often include secondary or modifier specifications, such as thermal stability, thermal sensitivity shift and thermal zero shift, that account for the temperature effects on MTE performance. These secondary tolerance limits must be added to the baseline tolerance limits if the MTE is to be used outside the associated baseline temperature range. Temperature effects specifications are typically stated as a percent of full scale per degree Fahrenheit or Celsius (i.e., %FS/°C).

When MTE performance is influenced by mechanical conditions, additional specifications for acceleration, vibration and acoustic noise, and shock are often included. Acceleration sensitivity specifications provide an indication of the effect of acceleration on MTE performance. Acceleration sensitivity specifications are typically stated as a percent of full scale per unit acceleration (i.e., %FS/g).

Vibration and acoustic noise specifications provide an indication of the error introduced when the MTE is subjected to vibration levels of specified amplitude and frequency. Vibration sensitivity specifications are typically stated as a percent of full scale per unit acceleration (i.e., %FS/g). Acoustic noise specifications are typically stated as a percentage of full scale or in dB.

Mechanical shock resistance specifications typically indicate the maximum allowable sudden acceleration or deceleration that the MTE can endure without any significant impact on performance or potential damage. Shock resistance specifications are typically stated as a fixed force level for a given elapsed time (e.g., 30g's for 11 milliseconds).

3.3.4 Temperature Compensation
Some MTE incorporate hardware and/or software to compensate or "correct" for the effects of temperature on performance. In this case, the compensated temperature range is specified. However, compensation methods cannot entirely eliminate error due to temperature effects. Therefore, manufacturers also specify a temperature coefficient or temperature correction error. Temperature coefficient specifications are typically stated as a percentage change or full scale per degree Fahrenheit or Celsius.

3.3.5 Excitation Voltage or Current
Other influence quantities that can affect MTE performance, such as power supply or warm-up requirements, are also specified, if applicable. For example, some transducers require an external voltage or current to generate the necessary output response. Excitation voltage specifications include the voltage level, accuracy and maximum current that the excitation voltage can source. Excitation current specifications include the current level, accuracy and the maximum load resistance that the excitation current can drive. The specification may also include a voltage or current drift rate with temperature.

CHAPTER 4: ESTABLISHING PERFORMANCE CHARACTERISTICS

MTE performance parameters and attributes are often initially established from experimental data (i.e., measurements) gathered during the research and development (R&D) phase. In some cases, performance parameters may be derived from engineering analysis of the components or materials used for MTE development and production.

Whether MTE parameters and attributes are determined via measurement or engineering analysis, the R&D, manufacturing and marketing personnel must ultimately agree on which performance characteristics will be included in the final, published specifications.

To maintain product quality control, MTE manufacturers may adjust performance parameters to account for uncertainty in their measurement or analysis processes. The resulting specifications are a set of tolerance limits which the manufacturer guarantees (or warranties) the MTE performance will meet or exceed with some level of confidence.[5]

4.1   Engineering Analysis

In an engineering analysis, the following technical information are used to establish the characteristics that will determine whether or not the MTE will meet performance, quality and reliability requirements over the expected life and operating environments:

- Design drawings

- Material and component specifications

- Performance characteristics of similar or related MTE designs

Engineering analysis is typically employed in lieu of, or in addition to, testing and involves the application of mathematical or numerical modeling methods that relate MTE inputs and outputs. These models must have sufficient detail to accommodate performance prediction, uncertainty estimation, failure or degradation mechanisms, and environmental effects. These analytical tools can be used to identify operational characteristics and determine the effects of various environmental conditions on performance.

A comprehensive engineering analysis can verify that the MTE parameters and attributes meet the basic functional requirements and that they meet the initial design tolerances. Engineering analysis can also identify potential design flaws and determine MTE parameters and attributes that must be established via testing.

4.1.1   Analysis Methods and Tools

Various analytical tools can be employed to assess which parameters significantly affect MTE performance. These tools can also be used to identify possible hardware and firmware deficiencies during MTE design and development.

4.1.1.1 Similarity Analysis

---

[5] Developing tolerance limits for MTE parameters and attributes is discussed in Chapter 5.

Verification by similarity is used to assess whether the MTE under evaluation is similar or identical in design, manufacturing process, and application to an existing device that has already been qualified to equivalent or more stringent performance criteria.

This method consists of reviewing and comparing the configuration, application, and performance requirements for similar equipment to the new MTE design. Prior test data, conditions and procedures for the similar equipment are also reviewed.

It is important that the application and operating environment of the new and existing devices or equipment are demonstrably similar. Any significant differences in the configuration, application or operating conditions indicate the need for further analysis and/or testing.

4.1.1.2 Failure Mode and Effects Analysis

Failure mode and effects analysis (FMEA) was initially developed in the 1940's for U.S. military applications. FMEA is currently used by a variety of industries to identify potential failure modes for products and to assess the consequences of these failures on functionality and performance.

FMEA also includes methods for assessing the risk associated with the failure modes, ranking them by the severity of the effect on functionality or performance, and identifying and implementing corrective actions on a prioritized basis. The two most common methods are the Risk Priority Numbers (RPN) method and the Criticality Analysis method.

The RPN method employs the following steps:

1. Determine all failure modes based on the functional requirements and their effects. A failure mode is the manner that a malfunction occurs in the MTE software, firmware or hardware part, component or subsystem. Examples of failure modes include, but are not limited to, electrical short-circuiting, corrosion or deformation. Examples of failure effects include, but are not limited to, degraded performance or noise. Each effect is given a severity number (S) from 1 to 10.

2. Identify and document all potential root causes or mechanisms for a failure mode. Examples of causes are: excessive voltage supply or operating temperature. The rate at which a failure occurs can be estimated by reviewing documented failures of similar products. Each failure mode is given a probability or occurrence number (O), from 1 to 10.

3. Estimate the likelihood of detecting the cause of each failure mode. A detection number (D), from 1 to 10, is assigned that represents the ability of removing defects or detecting failure modes during planned tests and inspections.

A list of possible failure mode occurrence, severity and detection rankings are shown in Table 4-1. The RPN is computed by multiplying the detectability, severity and occurrence.

$$RPN = D \times S \times O \tag{4-1}$$

The highest $RPN = 10 \times 10 \times 10 = 1000$ reflects that the failure mode is not detectable by inspection, the effects are very severe and there is a very high probability of occurrence. If the occurrence is very rare, then $O = 1$ and the RPN would decrease to 100.

Once the *RPN*s have been computed for the device, they can be used to determine the areas of greatest concern and to prioritize the failure modes that require corrective action. Failure modes with the highest *RPN*, rather than the highest severity, should be given priority for corrective action. These actions can include specific inspection, testing or quality procedures, complete redesign or selection of new components, adding redundant or backup components, and limiting environmental stresses or operating range.

Table 4-1. Examples of Failure Mode Occurrence, Severity and Detection Rankings

Ranking	Failure Rate (*O*)	Failure Effect (*S*)	Failure Detection (*D*)
1	1 in 1,000,000	None	Almost Certain
2	1 in 500,000	Very Low	Very High
3	1 in 100,000	Low	High
4	1 in 50,000	Low to Moderate	Moderately High
5	1 in 10,000	Moderate	Moderate
6	1 in 5,000	Moderate to High	Low
7	1 in 1,000	High	Very Low
8	1 in 500	Very High	Remote
9	1 in 100	Hazard	Very Remote
10	1 in 10	Hazard	Almost Impossible

The Criticality Analysis method ranks the significance of each potential failure mode for each part, component or subsystem of the design based on failure rate and severity. Criticality Analysis is used to prioritize and minimize the effects of critical failures early in the design phase.

In Criticality Analysis, each potential failure mode is ranked according to the combined influence of severity and probability of occurrence. To use the criticality analysis method, the following steps must be completed:

1. Define the reliability for each item (i.e., part, component or subsystem) and use it to estimate the expected number of failures at a given operating time.

2. Identify the portion of the item's reliability (or lack of thereof) that can be attributed to each potential failure mode.

3. Rate the probability of loss (or severity) that will result from each failure mode that may occur.

4. Calculate the failure mode criticality number (C_m) for each potential failure mode.

$$C_m = \beta \alpha \lambda_p t \qquad (4\text{-}2)$$

where

β = conditional probability of failure effect
α = failure mode ratio
λ_P = part failure rate per million hours
t = duration of required function expressed in hours or

17

operating cycles

5. Calculate the criticality number for each item for each severity category. This number is determined by adding all of the failure mode criticality numbers of the entire device or assembly with the same severity level.

$$C_r = \sum_{i=1}^{n} (C_m)_i \tag{4-3}$$

where n is the number of failure modes at the particular severity category.

FMEA or FMECA (Failure Mode, Effects and Criticality Analysis) is typically performed during product design and development, where initial performance requirements are established to minimize the likelihood of failures. At this stage, any design characteristics that contribute to possible failures are identified and mitigated or eliminated. This early identification and elimination of potential failure modes can significantly reduce future product redesign, rework, or recall.

4.1.1.3 <u>Sensitivity Analysis</u>
Sensitivity analysis involves performing numerical experiments in which the input parameters of the mathematical or computer model are systematically changed and the resulting output response analyzed. Consequently, sensitivity analysis can be used to identify which MTE parameters have the largest or smallest affect on overall performance.

The sensitivity analysis can be repeated for any number of individual input parameters. The number of numerical test runs increases with the number of values used for each input parameter. For example, if minimum and maximum values are examined for six input parameters, then all possible combinations of input conditions would require $2^6 = 64$ analysis runs.

As with physical experiments, one should strive to obtain the required input-output response information with a minimum number of test points. Design of experiments (DOE) techniques can be used to develop an optimum sensitivity analysis plan. DOE methods are discussed in Section 4.2.3.

There are two strategies for changing input parameter values.

1. All parameters are changed by the same percentage relative to their respective nominal value.

2. Each parameter is changed by the standard deviation of the corresponding probability distribution.

The second approach produces a more complex analysis because the individual input parameters are changed by different percentages. However, this approach may provide a more realistic portrayal of the actual input parameter variability.

There are also two methods for assessing the influence or relative importance of each input parameter on the model output.

18

1. Single-parameter Sensitivity Analysis. In single-parameter sensitivity analysis, the input parameters are assumed to be independent of one another and only one parameter is changed at a time, while the remaining parameters are kept at their nominal or base-case values.

   A variation of this method involves taking first-order partial derivatives of the model equation with respect to each parameter. The resulting sensitivity coefficient equations are then used to compute the relative sensitivities at nominal conditions.

2. Multi-parameter Sensitivity Analysis. In a multi-parameter sensitivity analysis, input parameters are assigned probability distributions that are propagated through the model using Monte Carlo simulation. Multi-parameter sensitivity analysis is a more generalized approach that accounts for parameter interactions.

The change in the model output response to low (L) and high (H) values for each parameter is ranked in order from large to small, as shown in Figure 4-1.

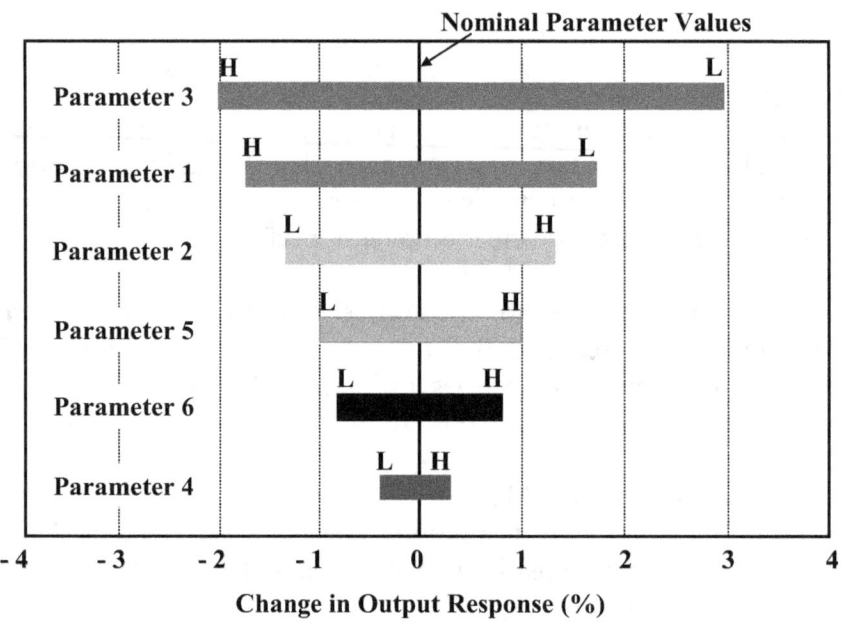

Figure 4-1 Sensitivity Diagram

Sensitivity analysis can be performed in the design, development and fabrication phases and is especially useful in evaluating the effects of environmental operating conditions and component wear or age on MTE performance.

4.2 First Article Testing

Testing of prototypes[6] and first production articles is often performed in the design and development phase to establish basic functionality, characterize performance and identify and mitigate any deficiencies. First article testing also provides an important assessment of the manufacturing process, equipment, and procedures before entering into the full production phase.

---

[6] Prototypes constitute items built during initial product design and development that may have partial or full functionality compared to the final realized product.

First article tests, also known as pre-production qualification tests, are conducted on a single item or on a sample taken at random from the first production lot. Prototype and first article tests should provide an explicit measure of MTE performance parameters and attributes during exposure to applicable operating conditions and environments. At a minimum, the tests must provide sufficient information to determine that the device performs satisfactorily for its intended application.

Methods and procedures for testing and reporting the functional and performance parameters for a variety of measuring equipment and devices can be obtained from professional organizations such as ISA, ASTM, ASME and IEC.

For example, ISA-37.10-1982 (R1995) *Specifications and Tests for Piezoelectric Pressure and Sound Pressure Transducers*, provides standardized methods and procedures for testing and reporting the design and performance parameters of piezoelectric pressure and sound-pressure transducers. Standards for the testing and reporting of MTE performance parameters are listed in Appendix B.

When practical, it is also a good practice to test the individual materials, parts, components, subassemblies and interfaces of the MTE to help identify the parameters and attributes that are critical to achieving the overall performance requirements established during design and development.

The objectives of the test program should be to

- Establish baseline performance parameters and attributes (e.g., sensitivity, linearity, zero offset)
- Quantify the effects of environmental conditions (e.g., temperature, vibration, humidity) on these performance parameters and attributes.
- Establish environmental operating limits (e.g., -20 to 70 °C, 0 to 50% RH)
- Identify other conditions that may influence these characteristics or parameters (e.g., external power supply).

To achieve these objectives, the test program should include both functional and environmental testing scenarios that span nominal, typical and extreme conditions that could occur during actual or expected operation.

Functional tests are conducted at baseline or nominal conditions and are usually conducted before and after environmental tests. Environmental test are conducted at expected environmental operating conditions and may include subjecting materials, components and subassemblies to mechanical shock, vibration, and electromagnetic environments. Destructive testing and inspection of disassembled parts and components should be included in the test program, as appropriate, to identify failure modes and mechanisms.

The test program document should include:

- The nomenclature and identification of the test article material, component, subassembly or assembly.

- The performance parameters and attributes to be evaluated.

- The functional and environmental tests conducted.

- The control tolerance limits to be maintained for the environmental and other operating conditions.

- The identification of measuring, test and data acquisition equipment to be used.

- Schematics or diagrams showing the identification, location, and interconnection of test equipment, test articles, and measuring points.

- The sequence and procedural steps for conducting the tests.

4.2.1 Environmental Testing

Environmental tests are most commonly performed on equipment used in military, maritime, aviation and aerospace applications to verify that the selected materials, parts, components and finished products can withstand the rigors of harsh environments, such as

- Extreme temperature, pressure and humidity ranges

- Rapid temperature changes

- Vibration, acceleration, and mechanical shock

- Solar radiation and electromagnetic fields

At a minimum, environmental testing provides the manufacturer a means of determining how their product withstands environmental storage conditions and vibration stresses encountered during shipping. These tests should also establish the appropriate environmental operating ranges of the device. More importantly, however, environmental testing provides the data necessary to evaluate product reliability under conditions it may encounter over its lifetime.

Environmental test methods can include a single factor such as temperature or combined factors such as temperature, humidity and vibration. A few common environmental test methods are summarized in the following subsections. Environmental testing standards, handbooks and other useful documents are listed below.

- MIL-STD-810F, *Department of Defense Test Method Standard for Environmental Engineering Considerations and Laboratory Tests*, 2000.

- NASA SP-T-0023 Revision C, *Space Shuttle Specification Environmental Acceptance Testing*, 2001.

- MIL-HDBK-2036, *Department of Defense Handbook Preparation of Electronic Equipment Specifications*, 1999.

- MIL-HDBK-2164A, *Department of Defense Handbook Environmental Stress Screening Process for Electronic Equipment*, 1996.

- IEC 60068-1, *Environmental Testing Part 1: General and Guidance*, 1988.

4.2.1.1 Temperature Tests

Changes in environmental temperature can temporarily or permanently degrade the functionality and performance of a device by changing the properties or dimensions of the material(s) used in its manufacture. Temperature tests are typically conducted using control chambers, ovens or baths to evaluate the short-term and long-term effects on the stability of baseline performance parameters.

Temperature cycling tests subject the material, part, component or article to multiple cycles between predetermined extremes (e.g., - 50 to 50 °C). Temperature cycling is used to evaluate the effects of severe temperature conditions on the integrity and performance of the tested item and to identify failure modes resulting from fatiguing, thermal expansion and contraction, or other thermally induced stresses.

Exposure to rapid temperature changes from extreme cold to hot environments can result in failures or permanent changes in the electrical or mechanical characteristics of materials and components. Thermal shock tests can be conducted in a liquid medium, using hot and cold baths, or in an air using hot and cold chambers.

4.2.1.2 Humidity Tests
Humidity tests are typically run in environmental control cabinets or chambers to simulate real-world environments. Cyclic humidity tests are conducted to simulate exposure to high humidity and heat typical of tropical environments. Excess moisture is especially damaging to electronic equipment, resulting in the corrosion and/or oxidation of materials and components. Low humidity tests are run to evaluate the effects of extremely dry climates. Low humidity can cause brittleness in materials and cause high electrostatic discharge conditions.

There are two types of high humidity tests: non-condensing and condensing. Non-condensing humidity tests are run at a constant temperature, with a high relative humidity, typically greater then 95%. Condensing humidity tests consist of temperature cycling in high relative humidity air. The temperature cycling causes the moisture to condense on the test article surfaces. In some cases, the moisture laden air will also migrate inside the test article.

4.2.1.3 Low Pressure (Altitude) Tests
Testing for altitude allows a manufacturer to simulate the effects of low pressure and high altitude. Most environmental test chambers are capable of simulating altitudes up to 200,000 feet above sea level or pressures as low as 0.169 Torr. Altitude testing is especially important for devices and equipment used in defense and aerospace applications.

4.2.1.4 Vibration and Shock Tests
Manufacturers perform vibration and shock tests to determine if a product can withstand the mechanical stresses encountered during production, transport or end use. Vibration or shock induced failures typically result from accumulated fatigue damage during high stress cycles that occur at resonance response frequencies.

Shaker tables are often used to perform sinusoidal, random, and shock vibration tests. The random vibration stresses experienced in real-world environments are best simulated by multiple-axis simultaneous shaking of the test article.

4.2.2 Accelerated Life Testing
In many instances, it may be necessary to conduct tests to accelerate the aging of devices or equipment to identify problems that could detrimentally affect their reliability and useful life. Government agencies, such as the DOD, FAA and NASA require that instrument and equipment

manufacturers provide life expectancy or Mean Time to Failure (MTTF) data for their products.

Accelerated life testing (ALT) is often conducted when normal functional and environmental testing of large sample sizes yields few or no failures. ALT is conducted on a selected sample of parts, components or devices to force them to fail more quickly than they would under normal usage conditions. The primary purpose of ALT is to quantify the long-term reliability of the product. ALT also allows manufacturers to observe failures of their products to better understand these failure modes.

There are two categories of accelerated testing: qualitative and quantitative. Qualitative ALT focuses on identifying probable failure modes without attempting to make any predictions about the product's life expectancy under normal usage conditions. Quantitative ALT focuses on using the test data to predict the expected life or, more specifically, the MTTF and reliability of the product.

ALT is designed to apply functional and environmental stresses that exceed the typical limits or levels the product will encounter under normal usage conditions. The applied stresses can be constant, step-wise, ramped, cyclic, or random with respect to time. The appropriate stress application should be based on the frequency or likelihood of occurrence of the stress event.

Preferably, the stress application and limits used during ALT should be based on measurements of the actual usage conditions. If these stresses or limits are unknown, preliminary testing of small sample sizes can be performed using design of experiments methods to ascertain the appropriate stresses and stress limits.

4.2.2.1 Highly Accelerated Life Testing (HALT)
HALT is typically performed on prototypes or first articles so that design flaws or deficiencies can be identified and corrected prior to full scale production. The purpose of HALT to improve the product by focusing on corrective measures and improvement opportunities in the early stages of development. HALT uses step-by-step cycling of environmental conditions such as temperature, shock and vibration to induce failure modes. HALT often also includes the simultaneous cycling of environmental conditions, such as temperature and vibration, to provide a closer approximation of real-world operating conditions.

4.2.2.2 Highly Accelerated Stress Screening (HASS)
HASS applies stresses similar to those used in HALT, but without damaging the product. The production units are tested using combined environments, at lower levels and/or durations compared to those used for HALT testing. HASS is often used in an on-going screening process where tests are performed on actual production units to verify that they operate properly. HASS uses environmental variables such as fast temperature cycling, shock and multi-axis vibration as a means of screening production units to ensure that they perform reliably.

The HASS data are also mathematical modeled and analyzed to estimate of the product's life and long-term performance characteristics under normal usage and environmental conditions. The test results can also be used to establish the operational stress levels, within which the parameters and attributes of the device meets performance requirements.

4.2.3 Design of Experiments

In many instances, it is advantageous to incorporate statistical testing or design of experiments (DOE) methods to obtain maximum useful information while controlling costs. DOE is a scientific approach which allows the researcher or engineer to gather meaningful information to better understand a process or product and to determine how the input variables (factors) affect the output variables (responses).

Understanding which factors are critical to improving process efficiency or product performance can then be used to determine which factors should be controlled and establish acceptable tolerance limits. For example, the response in the output variable Y to a change in input factor X_1 may be minor; while a significant change in output response may result from a comparable change in input factor X_2, as shown in Figure 4-2.

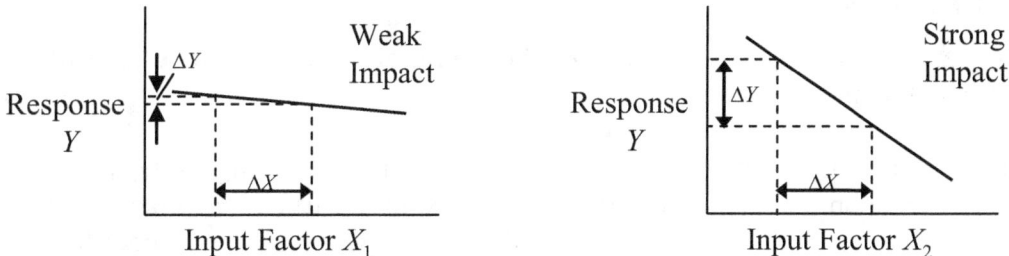

Figure 4-2 Response of Output Y to Input Factors X_1 and X_2

In this example, there is a clear benefit to limiting the input factor X_2 to a specific range to control the output response.

4.2.3.1 Classical Testing Strategy

The classical approach to DOE is to change one input factor at a times, holding all other input variables constant. This approach can be appealing in its simplicity. However, it can also lead to large numbers of tests run, depending upon the number of factors evaluated. For example, a classical testing strategy using four independent input factors, with three different values for each factor, would require a full factorial of 4^3 possible input variable combinations or 64 different tests. This does not include replication or repeat testing.

The classical testing strategy incorporates the following assumptions:

- The response (or change in output variable) is likely to be a complicated function of any change in input variable and, therefore, requires many experimental testing levels for each input variable.

- The interactions between input variables are negligible so that the effect of any other input variable is simply to raise or lower the function, but not to change its shape or slope.

- The errors associated with the testing and measurement process are negligible relative to the effects of changes in the input factors.

If these assumptions hold, then changing one factor at a time may be a good testing strategy. However, in reality, this testing strategy can be time consuming and costly. In addition, assuming negligible interactions between input variables may not necessarily provide a reliable or accurate depiction of the process or product being evaluated.

4.2.3.2 Statistical Testing Strategy

This DOE approach uses modified factorial designs to determine a statistically significant number of tests required to gain an understanding of the relationship between input factors and output responses.

The statistical approach incorporates the following assumptions:

- Most response functions for a given input factor are relatively smooth, with perhaps an upward or downward curvature, over the range of experiments.

- The slope or shape of a response function may change significantly as a result of changes in other input factors. Such interactions between input factors are considered typical.

- Testing and measurement errors may not be negligible relative to the effects of changes in the input factors.

If these assumptions hold, then the statistical approach is a good testing strategy. The number of test cases depends upon the number of input factors and the values of each factor. However, the number of test cases using a statistical design will be much less than the number for a comparable classical design.

There are a number of different statistical methods that can be used to design a test matrix. Some commonly used methods include:

- Two-level Factorial
- Plackett-Burman
- Three-level Factorial
- Box-Behnken

Two-level Factorial

Two-level factorial experimental design consists of 2^n distinct test cases, where n is the number of input factors to be evaluated at two levels each (low and high). For example, an experiment designed to evaluate three input factors (i.e., X_1, X_2, X_3) each having a low and high value would require $2^3 = 8$ test cases. If the three input factors were depicted in a three dimensional plot, then these test cases would, in effect, be the corners points of a cube, as shown in Figure 4-3.

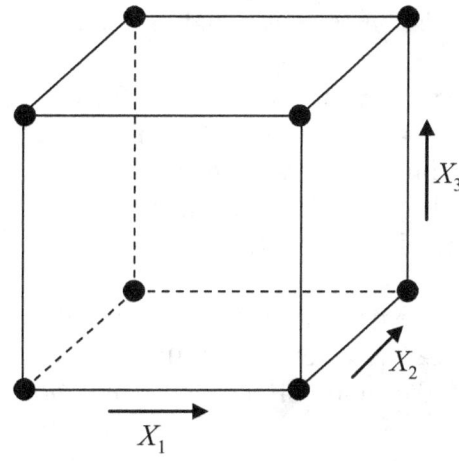

Figure 4-3 Two-level 2^3 Factorial Design

Each edge of the cube represents one factor in the experiment. Each test case represents a set of experimental conditions at which one or more experimental runs will be made. With more factors, the cube becomes what mathematicians call a hypercube.

The two-level factorial design can provide enough data to determine the main effect of each input factor plus the interactions of the factors in combination. By definition, two factors, say X_1 and X_2, are said to interact if the effect of X_1 is different at different levels of X_2, as shown in Figure 4-4.

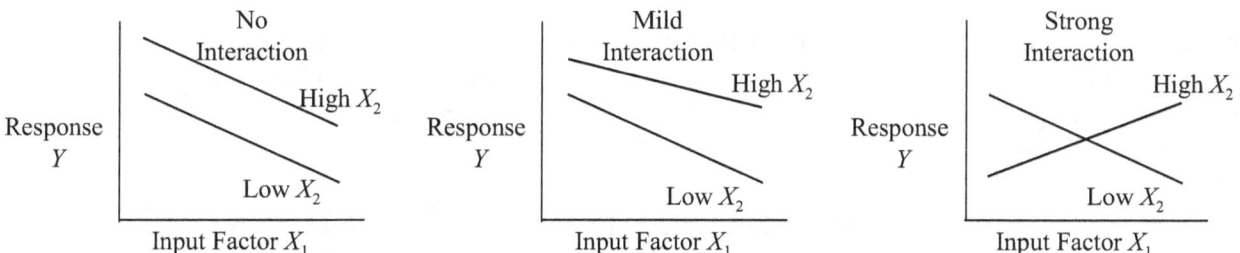

Figure 4-4 Interaction of Input Factors X_1 and X_2

Plackett-Burman
Plackett-Burman is a screening method that incorporates a subset of test cases generated from the two-level factorial 2^n design. This method is typically used when evaluating six or more input factors. Usually, in an overall experimental program, the application of a screening design comes first since, as its name implies, it is used to screen out the few really important factors with a minimum of testing.

The Plackett-Burman design is a specific fraction of the 2^n factorial that has properties that allow efficient estimation of the effects of the factors under study. Plackett-Burman designs are available from nearly every multiple of four to one hundred test cases or trials, the most useful ones are for 12, 20, or 28 trials that nominally handle up to 11, 19, and 27 factors, respectively.

The results of the screening tests may be directly applicable to the overall study if a significant impact in the response is observed at some experimental condition. More likely, however, once the important factors are identified from screening tests, then a more detailed set of experiments

26

can be conducted using a 2^n factorial where n is equal to the reduced number of input factors.

Response surface designs are used to estimate curvature effects. To do this, the experimental design must have at least three levels (low, mid, and high) for each independent factor, as shown in Figure 4-5. Using only low and high values for each input factor X would result in a linear curve fit. Using low, mid and high values for X would provide a good indication if the output Y responds in a non-linear fashion to changes in X.

Plackett-Burman designs can only be used to build very limited response surface models because they estimate the constant and linear terms only. Similarly, 2^n factorial designs support limited response surface models comprised of main effects and interactions.

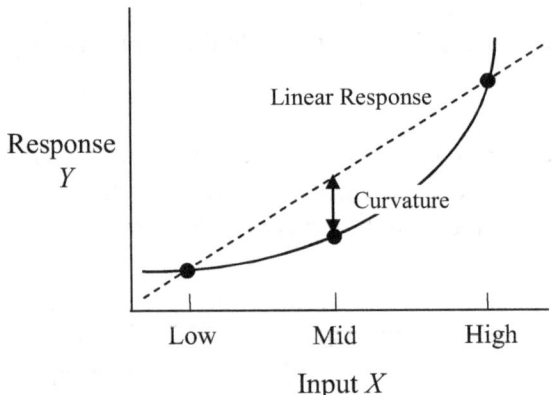

Figure 4-5 Midpoints used to Test for Curvature (non-linearity)

It may be feasible to use a full three-level or 3^n factorial design to provide estimates of linear, quadratic (curvature) and interaction effects. However, a disadvantage of using a 3^n factorial design is the large number of experimental trials or tests required, as shown in Table 4-2.

Table 4-2. Tests Required for Three-Level Factorial Designs

Number of Input Factors	Number of 3^n Factorial Tests
2	9
3	27
4	81
5	243
6	729

Box-Behnken

The Box-Behnken design employs a subset of the full 3^n factorial design. For example, consider an experiment designed to evaluate three input factors (X_1, X_2, X_3) each having a low, mid and high value. If the three input factors were depicted in a three dimensional plot, then Box-Behnken test cases would be located at the mid points of the edges and the center of the cube, as shown in Figure 4-6.

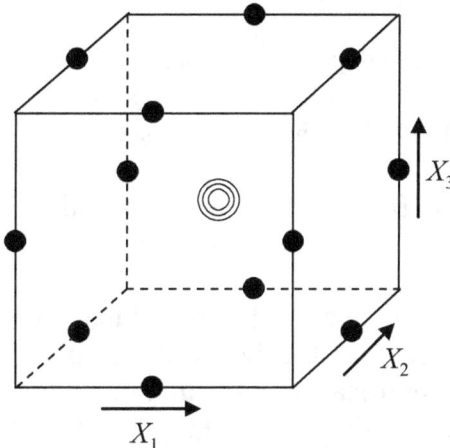

Figure 4-6 Box-Behnken Three-level, Three-Factor Design

The three replicated center points depicted in Figure 4-6 have two functions. First, they provide a measure of experimental repeatability. Second, they are sufficient in number to give a relatively consistent prediction of output response as a function of distance from the center or nominal conditions within the design region.

Therefore, a Box-Behnken design for a three-level, three-factor matrix would require only 15 test cases compared to 27 test cases for the full 3^3 factorial (excluding replications at the nominal or mid-point condition). Similarly, a Box-Behnken design for a three-level, four-factor design would require only 27 test cases compared to 81 test cases for the full 3^4 factorial design.

4.2.4   Data Analysis Methods and Techniques

Practical knowledge about the materials, parts and components and the scientific principles of operation should be the basis for analyzing the data obtained from prototype or first article tests. When a mathematical model cannot be developed from scientific principles, then an empirical model is developed from statistical analysis of the test data.

Empirical modeling provides a functional equation that relates changes in a given output response to changes in one or more factors, captures data trends and can be used to predict behavior within the experimental data domain.[7] Three common statistical data modeling techniques include:

- Regression Analysis
- Analysis of Variance
- Response Surface Analysis

4.2.4.1 <u>Regression Analysis</u>

Regression analysis is a statistical technique used for the modeling and analysis of data consisting of values of a dependent variable (response variable) and corresponding values of one or more independent variables (factors). The regression equation expresses the dependent variable as a function of the independent variables, corresponding coefficients and a constant or intercept term. The regression coefficients are estimated to achieve a "best fit" of the data using the least squares method.

---

[7] Empirical models are not considered useful or reliable tools for extrapolating beyond the realm of actual test data.

The linear regression equation for the dependent variable, Y, takes the form

$$Y = b_1X_1 + b_2X_2 + \ldots + b_nX_n + c \qquad (4\text{-}4)$$

where the b's are the regression coefficients for the corresponding X independent variables and c is the constant or intercept.

The regression coefficients represent the average amount that the dependent variable changes when the corresponding independent variable increases one unit and the other independent variables are held constant. The significance of the individual regression coefficients are assessed using t-tests. The regression line intercept represents the value of the dependent variable when all the independent variables are equal to zero.

Linear regression models like that shown in equation (4-4) do not contain interaction effects between the independent variables. Consequently, linear regression models are often referred to as main effects models. A main effect is the overall effect of each independent variable or factor by itself. As shown in Figure 4-4, an interaction occurs when the effects of one factor change depending on the value of another factor. Both main affects and interactions between factors can be accommodated using Analysis of Variance or Response Surface Analysis.

4.2.4.2 Analysis of Variance

Analysis of Variance (ANOVA) is a statistical process for evaluating the concurrent effects of multiple independent variables or factors on one dependent variable. ANOVA also allows the evaluation of interaction and main effects. In ANOVA, the relative importance of various combinations of independent variables are assessed using F-tests.

Although the effects of a single factor can be evaluated, ANOVA is typically applied when studying the effects of two or more factors. The term "way" or factor is often used to describe the number of independent variables evaluated. For example, one-way or one-factor ANOVA measures the effect of one independent variable on a dependent variable, and two-way or two-factor ANOVA measures the effect of two independent variables on the dependent variable.

When conducting ANOVA, the variation of each factor should be roughly equal to avoid difficulties with the probabilities associated with the significance tests. Ideally, the factors should also have the same sample size or number of observations.[8] Otherwise, the ANOVA can become a very complicated endeavor.

An ANOVA with two factors, X_1 and X_2, can be written as

$$Y = b_0 + b_1X_1 + b_2X_2 + b_{12}X_1X_2 \qquad (4\text{-}5)$$

where Y is the response for given levels of the factors X_1 and X_2 and the X_1X_2 term accounts for a possible interaction effect between X_1 and X_2. The constant b_0 is the response of Y when both main effects equal zero.

---

[8] An experimental design where all cells (i.e. factorial combinations) have the same number of observations or replications are called *balanced factorial designs*.

An ANOVA with three factors X_1, X_2 and X_3 would be written as

$$Y = b_0 + b_1 X_1 + b_2 X_2 + b_3 X_3 + b_{12} X_1 X_2$$
$$+ b_{13} X_1 X_3 + b_{23} X_2 X_3 + b_{123} X_1 X_2 X_3 \qquad (4\text{-}6)$$

where the main effects and two-factor interaction terms are accompanied by a three-factor interaction term $X_1 X_2 X_3$. As in equation (4-5), the constant b_0 is the response of Y when the main effects equal zero.

In ANOVA, a variety of null hypotheses are tested for the mean or average responses when the factors are varied. For the two-factor ANOVA, the possible null hypotheses are:

1. There is no difference in the means of factor X_1
2. There is no difference in the means of factor X_2
3. There is no interaction between factors X_1 and X_2

The purpose of the hypotheses tests is to determine whether the different levels of factor X_1, or factor X_2, really make a difference in the output response, and whether the $X_1 X_2$ interaction is significant.

For example, consider the analysis of a two-factorial experiment in which factor X_1 has r levels, factor X_2 has c levels and each test condition (cell) is replicated n times. In this case, the total number of test conditions (or cells) is $r \times c$ and total number of data points is $r \times c \times n$. The resulting response data would be similar to that shown in Table 4-3.

Table 4-3. ANOVA Data for Two-Factorial Experiment

		Factor X_2			
		1	2	L	c
Factor X_1	1	$Y_{111}, Y_{112}, L, Y_{11n}$	$Y_{121}, Y_{122}, L, Y_{12n}$	L	$Y_{1c1}, Y_{1c1}, L, Y_{1cn}$
	2	$Y_{211}, Y_{212}, L, Y_{21n}$	$Y_{221}, Y_{221}, L, Y_{22n}$	L	$Y_{2c1}, Y_{2c2}, L, Y_{2cn}$
	Ν	Ν	Ν	Ν	Ν
	r	$Y_{r11}, Y_{r12}, L, Y_{r1n}$	$Y_{r21}, Y_{r22}, L, Y_{r2n}$	L	$Y_{rc1}, Y_{rc2}, L, Y_{rcn}$

The null hypotheses for the main effects and interaction effects are tested by calculating a series of sums of squares (SS) that are the foundation of the variance (σ^2) estimates. The equations for performing an ANOVA for this data are given in Table 4-4.

Table 4-4. General Two-Factor ANOVA Equations

Source of Variation	SS	df	MSQ (est. of σ^2)	F Ratio
Rows (X_1)	$c \times n \left[\sum_{i=1}^{r} \left(\overline{Y}_{ri} - \overline{\overline{Y}} \right)^2 \right]$	$r - 1$	$\dfrac{SS_{X_1}}{df_{X_1}}$	$F_{X_1} = \dfrac{MSQ_{X_1}}{MSQ_W}$

Source	Sum of Squares	df	MSQ	F Ratio
Columns (X_2)	$r \times n \left[\sum\limits_{j=1}^{c} \left(\bar{Y}_{cj} - \bar{\bar{Y}} \right)^2 \right]$	$c - 1$	$\dfrac{SS_{X_2}}{df_{X_2}}$	$F_{X_2} = \dfrac{MSQ_{X_2}}{MSQ_W}$
Interactions (I)	$n \left[\sum\limits_{i=1}^{r} \sum\limits_{j=1}^{c} \left(\bar{Y}_{ij} - \bar{Y}_{ri} - \bar{Y}_{cj} - \bar{\bar{Y}} \right)^2 \right]$	$(r - 1)(c - 1)$	$\dfrac{SS_I}{df_I}$	$F_I = \dfrac{MSQ_I}{MSQ_W}$
Within Groups or Cells (W)	$\sum\limits_{i=1}^{r} \sum\limits_{j=1}^{c} \sum\limits_{k=1}^{n} \left(Y_{ijk} - \bar{Y}_{ij} \right)^2$	$(r \times c)(n - 1)$	$\dfrac{SS_W}{df_W}$	
Total	$\sum\limits_{i=1}^{r} \sum\limits_{j=1}^{c} \sum\limits_{k=1}^{n} \left(\bar{Y}_{ijk} - \bar{\bar{Y}} \right)^2$	$r \times c \times n - 1$		

The row sum of squares compares each row mean, $\bar{Y}_{ri}$, with the overall or grand mean of the data, $\bar{\bar{Y}}$. The variance for each row of data is the sum of the mean squares (MSQ) divided by the associated degrees of freedom (df). The within-group sum of squares divided by the associated degrees of freedom produces an unbiased estimate of σ^2, regardless of whether any null hypothesis is true. This occurs because each individual observation within a cell, Y_{ij}, is compared with its own cell mean value, $\bar{Y}_{ij}$. This is why the within-group mean square (MSQ_W) is used as the denominator of the F ratio equations.

If the three null hypotheses are true, then the F values constitute ratios of two estimates of the same population variance, σ^2, and the F ratios would be small. Conversely, if the three null hypotheses are false, then the estimate of σ^2 will result in large values of F and the corresponding null hypotheses would be rejected.

The rejection criteria for the F-tests are established using tabulated values of the F distribution for a specified significance level (e.g., 0.05), the degrees of freedom for the numerator and the degrees of freedom for the denominator in the F ratio equation. If calculated F ratio exceeds the corresponding value obtained from the F distribution table, then null hypothesis can be rejected at the specified significance level.

If the analysis fails to reject the null hypothesis of no interaction between X_1 and X_2, then the interaction term can be removed from Table 4-4. The remaining two hypotheses for the effects of X_1 and X_2 are then re-evaluated assuming that no interaction exists. The procedure is to combine the interaction sum of squares and degrees of freedom with the within-group or within-cell values.

Any variation previously attributed to possible interaction effects is now identified as random fluctuations in the data. Conducting this second, revised ANOVA provides more powerful hypothesis tests for the main effects.

The procedural steps for the two-factor ANOVA test are summarized below.

1. State the null hypotheses.
2. Make the necessary assumptions (normal populations, equal variances, independent observations).

3. Collect experimental data from each of the populations under investigation.

4. Assuming that the null hypotheses are true, estimate the common variance (σ^2) of the populations using the four different methods listed in Table 4-4.

5. Compute the three F ratios using the equations listed in Table 4-4.

6. Obtain tabulated F values from the F distribution table, using the appropriate degrees of freedom and assuming some level of confidence.

7. Reject or fail to reject each of the null hypotheses based on whether or not the calculated F ratios exceed the tabulated F values.

8. (Optional) If the null hypothesis regarding interaction cannot be rejected, recalculate the ANOVA results by combining the interaction sum of squares and degrees of freedom with the within-cell values.

4.2.4.3 Response Surface Modeling

The application of ANOVA becomes untenable for data obtained from experimental studies of four or more factors. In such cases, response surface analysis is used to develop the empirical models. Response surface models can account for main effects and interaction effects. They can also contain quadratic or cubic terms to account for curvature (i.e., nonlinearity) effects.

A mathematical equation that is continuous and has derivatives of first order and higher terms in the interval being considered can be approximated by a Taylor series. This holds regardless of the complexity of the equation. However, very complex equations may require many terms in the Taylor series.

Polynomial approximations are commonly used to develop response curve models from test data. The mathematical basis for this is that a polynomial has the same form as a Taylor series expansion that has been truncated after a specific number of terms. Polynomial models are also useful because they can be readily differentiated and integrated.

A second order polynomial or quadratic model for a three-factor experimental design is

$$
\begin{aligned}
Y = b_0 &+ b_1 X_1 + b_2 X_2 + b_3 X_3 + b_{12} X_1 X_2 \\
&+ b_{13} X_1 X_3 + b_{23} X_2 X_3 + b_{11} X_1^2 + b_{22} X_2^2 + b_{33} X_3^2
\end{aligned}
\tag{4-7}
$$

where Y is the response for given levels of the factors X_1, X_2 and X_3. The two-factor interactions are $X_1 X_2$, $X_1 X_3$, and $X_2 X_3$. The constant b_0 is the response of Y when the main effects equal zero.

A third order polynomial or cubic model for the three-factor experimental data is given in equation (4-8).

$$
\begin{aligned}
Y = \text{quadratic model} &+ b_{123} X_1 X_2 X_3 + b_{112} X_1^2 X_2 + b_{113} X_1^2 X_3 \\
&+ b_{122} X_1 X_2^2 + b_{133} X_1 X_3^2 + b_{223} X_2^2 X_3 + b_{233} X_2 X_3^2 \\
&+ b_{111} X_1^3 + b_{222} X_2^3 + b_{333} X_3^3
\end{aligned}
\tag{4-8}
$$

32

Equations (4-7) and (4-8) are full response surface models, containing all possible terms. The linear terms produce a multidimensional response surface or hyperplane. The addition of interaction terms allows for warping of the hyperplane, while addition of the second order terms produce a surface with a maximum or minimum response. The addition of third and higher order terms introduce additional inflection points in the hyperplane.

It is important to note that, not of all of these terms may be required in a given application. When choosing a polynomial model, one must consider the trade-offs between developing a simplified description of overall data trends and creating a more robust prediction tool. Ultimately, the resulting model has to have a requisite "goodness of fit" in order to determine the most significant factors or optimal contours of the response surface.

In some product or process studies, there may be an interest or need to model the response of more than one output quantity to several input factors. This relationship can be expressed by one or more equations.

$$Y_1 = f_1\ (X_1, X_2, \ldots, X_n)$$
$$Y_2 = f_2\ (X_1, X_2, \ldots, X_n)$$
$$\mathsf{N}$$
$$Y_m = f_m\ (X_1, X_2, \ldots, X_n)$$

Multivariate response surface models are obviously much more complex, requiring the simultaneous solution of the multiple polynomial equations. To do this, the system of equations are formulated in matrix notation

$$\boldsymbol{Y = Xc} \tag{4-9}$$

where $\boldsymbol{Y}$ is the vector of m observed response values $[Y_1, Y_2, \ldots, Y_m]$, $\boldsymbol{c}$ is the vector of n unknown coefficients $[c_0, c_1, \ldots, c_{n-1}]$, and $\boldsymbol{X}$ is the $m \times n$ matrix formed by input factor terms.

$$\boldsymbol{X} = \begin{bmatrix} 1 & X_1 & X_2 & X_1X_2 & \mathrm{L} & X_n^2 \\ \mathrm{M} & \mathrm{M} & \mathrm{M} & X_1X_2 & \mathrm{O} & \mathrm{M} \\ 1 & X_1 & X_2 & X_1X_2 & \mathrm{L} & X_n^2 \end{bmatrix} \tag{4-10}$$

Equation (4-9) is solved either by direct computational methods such as Gaussian elimination or by iterative methods such as Gauss-Seidel.[9]

4.2.4.4 Graphical Techniques
There are an abundance of statistical graphical techniques that can be used for qualitative data analysis. Graphical techniques include, but are not limited to:

- Scatter plots
- Box plots
- Block plots
- Youden plots

---

[9] See, for example, *Numerical Recipes in Fortran*, 2nd Edition, Cambridge University Press.

- Star plots

These graphical techniques are useful for identifying trends, detecting outlier data and determining factor effects. Descriptions and illustrative examples for various graphical data analysis techniques can be found in the NIST/SEMATEC: *e-Handbook of Statistical Methods*.

4.3   Measurement Quality Assurance

Measurement quality assurance methods and procedures should be implemented to ensure that the testing program results in accurate measurements. To achieve this, both quality control and assessment activities should be developed and implemented. These activities include

- The implementation of a documented process for the periodic calibration of test equipment to verify that they are capable of accurately and reliably performing their intended measurement tasks.

- An assessment of measurement uncertainty for all equipment calibrations using appropriate analysis methods and procedures that account for all sources of uncertainty (e.g., biases of reference standards or materials, display resolution, environmental conditions).

ISO/IEC 17025:2005 *General Requirements for the Competence of Testing and Calibration Laboratories* is a consensus standard that contains the requirements that testing and calibration laboratories must meet to demonstrate that they operate a quality program or system and are capable of producing technically valid results.

4.3.1   Test Equipment Calibration

All equipment and devices used to obtain measurements must be properly calibrated to establish and maintain acceptable performance during testing. ANSI/NCSLI Z540.3-2006 *Requirements for the Calibration of Measuring and Test Equipment* is the U.S. consensus standard that establishes the technical requirements for managing MTE and assuring that their calibrated parameters conform to specified performance requirements.

ISO/IEC 17025 and ANSI/NCSLI Z540.3 also require that the measurement results obtained during equipment calibration must be traceable to a national measurement institute, such as the U.S. National Institute of Standards and Technology (NIST).[10]

The interval or period of time between calibrations may vary for each measuring device depending upon the stability, application and degree of use.[11] In some cases, it may be necessary to conduct pre-test and post-test calibrations to ensure proper equipment performance during testing and reduce the need for costly retesting.

4.3.2   Measurement Uncertainty

All measurements are accompanied by errors associated with the measurement equipment used, the environmental conditions during measurement, and the procedures used to obtain the measurement. Therefore, test data cannot simply be used to characterize MTE performance without an understanding of the uncertainty that these measurement errors introduce. Similarly,

---

[10] Measurement traceability is discussed further in Section 10.4.4.

[11] Periodic calibration is discussed in Section 10.4.

calibration data alone cannot be used to verify conformance of test equipment to manufacturer specifications without increasing false accept or false reject risk.[12]

Our lack of knowledge about the sign and magnitude of measurement error is called measurement uncertainty. Therefore, the data obtained from testing and calibration are not considered complete without statements of the associated measurement uncertainty.

Analytical and empirical models are used to conduct numerical experiments. These numerical experiments employ underlying assumptions and input conditions that have associated uncertainties. Therefore, these engineering studies should also include the analysis and reporting of uncertainty estimates.

Since uncertainty estimates are used to support decisions, they should realistically reflect the measurement or analytical modeling process. In this regard, the person tasked with conducting an uncertainty analysis must be knowledgeable about the process under investigation.

An in-depth coverage of key aspects of measurement uncertainty analysis and detailed procedures needed for developing such estimates is provided in NASA *Measurement Quality Assurance Handbook*, Annex 3 – *Measurement Uncertainty Analysis Principles and Methods*.

4.4   Engineering and Testing Reports

The engineering analysis or testing report is the primary outcome of the modeling or experimental study. These technical reports will be used as the basis of future decisions and provide valuable information for possible future studies.

Therefore, the report should be comprehensive: clearly defining the objectives of the study, describing the testing and/or analysis methods used to achieve these objectives, and concisely presenting the analysis or test results.

4.4.1   Reporting Engineering Analysis Results

The engineering report should present the analysis results and conclusions, the technical concepts and methodology employed and the test data evaluated in a logical, concise manner. The report should also include a graphical depiction of the prototype or first article design. The report contents should include:

- An executive summary that highlights the main objectives of the analysis, the methodology used and the main findings and conclusions.

- An introduction explaining the purpose and scope of the study, background information, schematics and drawings, and a review of literature or other resources used in the study.

- A description of the methodology, mathematical approach or functional relationships used, any underlying assumptions or simplifications applied, and how the data/information were gathered, generated and analyzed.

- A discussion of the analysis results highlighting expected and unexpected findings, supported by graphs, tables and charts, and a comparison to other related

---

[12] The application of uncertainty estimates for assessing the risk of falsely accepting or rejecting a calibrated parameter is discussed in Sections 10.4.5 and 11.3.

engineering studies or analyses, if applicable.

- Conclusions summarizing what was learned, what remains to be learned (if applicable), any limitations of the analysis and recommendations for future analysis or testing, if needed.

- Appendices containing raw data, sample calculations, mathematical algorithms and other supplemental information.

4.4.2   Reporting Test Results

An essential outcome of any experimental testing program should be a comprehensive report that clearly defines the test objectives, identifies and describes the configuration of the item(s) tested, documents the test plan, and specifies the environmental test conditions and procedures used to achieve these objectives.

The testing report is similar in content and layout to an engineering analysis report and should include the following:

- An executive summary that highlights the main objectives of the experiments, the experimental design approach and analytical methods used and the main findings and conclusions.

- An introduction explaining the purpose and scope of the experimental study and providing background information, schematics and drawings, and a review of literature or other resources.

- A description of the experimental design approach and supporting diagrams, schematics or drawings depicting the test article(s), the overall test setup and the specific test equipment or reference standards used.

- A discussion of the data analysis methods used, along with any underlying assumptions or simplifications applied, and a summary of the data analysis results highlighting expected and unexpected findings, supported by graphs, tables and charts, and a comparison to other related experimental studies or engineering analyses, if applicable.

- Additional discussions identifying and describing any detected deficiencies, functional failures or unsatisfactory operations in the article(s) encountered during testing, along with recommendations for corrective actions or discussion of corrective actions that have been implemented or are pending.

- Conclusions summarizing the overall findings, including the recommended tolerance limits within which each tested parameter or attribute is considered to provide acceptable performance, and any recommendations for future testing and/or analysis, if needed.

CHAPTER 5: DEVELOPING SPECIFICATIONS

Specifications should realistically reflect MTE performance under clearly defined operating conditions. MTE manufacturers typically establish baseline performance parameters such as repeatability, linearity, hysteresis, response time, span, and threshold for limited operating conditions. Electronic equipment, for example, are typically evaluated under environmental conditions of 23 °C ± 5 °C, with negligible exposure to mechanical shock, vibration and electromagnetic interference.

However, many MTE are designed and manufactured to perform reliably over an extended range of operating conditions. In these cases, it is important to characterize the change in MTE performance over an applicable range of environmental operating conditions. It is also important to develop specifications for the drift in MTE parameter performance over time.

Baseline, environmental and drift parameters are often initially established from experimental data (i.e., measurements) gathered during the research and development (R&D) phase. Environmental and drift specifications may also be determined from empirical data gathered during the product development and manufacturing phase. In some cases, the performance parameters may be derived from engineering analysis of the components or materials used to develop the product.

The results of testing and/or engineering analysis are used in the development of confidence or tolerance limits that establish the produced item is in accordance with its functional, performance, and design requirements. Ideally, specification limits should be established from the observed behavior for the population or sample of devices using DOE methods.

5.1   Engineering Analysis

As discussed in Chapter 4, mathematical or empirical models can be used to predict how a specified range of a given parameter or attribute affects overall performance. This information can then be compared to the initial performance criteria for the device and used to modify or adjust the original design tolerances. These analytical tools can also be used to establish additional modifier tolerance limits to account for the effects of variations in environmental and other operating conditions.

Absent mathematical or empirical models, the specification limits for the parameters and attributes must be developed by comparing the design, configuration, application and performance requirements of the new device to existing MTE. If they are demonstrably similar, then the performance specifications for the existing MTE can be used as a basis for establishing comparable specifications for the new device. However, if significant differences in the configuration, application or operating conditions of the new and existing MTE are identified, then new device will require testing and analysis.

5.2   Final Article Testing

Final article tests, also known as production qualification tests, are conducted on a single item or on a sample taken at random from a production lot. Final article testing and inspection should be performed in a manner similar to prototype or first article testing and under conditions that simulate end-use as closely as possible without damaging the item or product.

The objectives of the testing program should be to

- Provide a valid measure of performance, quality and reliability for important parameters and attributes identified during prototype or first article testing.

- Establish baseline tolerance limits for these parameters and attributes.

- Quantify the effects of environmental conditions (e.g., temperature, vibration, humidity) on these parameters and attributes.

- Establish environmental operating limits.

- Identify other operating conditions (e.g., external power supply) that may influence these parameters and attributes.

- Establish secondary or modifier tolerance limits (e.g., thermal zero shift, thermal sensitivity shift, vibration sensitivity).

To achieve these objectives, the test program should include both functional and environmental testing scenarios that span all conditions, including nominal, typical, and extreme, that could occur during actual or expected operation.[13]

The degree, duration and number of tests performed should be sufficient to provide assurance that the final article is inherently capable of meeting the established performance, quality and reliability requirements. Destructive testing and inspection of disassembled parts and components may be included in the test program as appropriate.

The test program document should include:

- The nomenclature and identification of the article (e.g., material, component, subassembly or assembly) to be tested.

- The performance parameters and attributes to be evaluated.

- The functional and environmental tests conducted.

- The control tolerances to be maintained for the environmental and other operating conditions.

- The identification of measuring equipment to be used during the article testing.

- Schematics or diagrams showing the identification, location, and interconnection of the test article, measuring equipment, and measuring points.

- The sequence and procedural steps for conducting the tests.

The data analysis methods should follow those used for prototype or first article testing described in Chapter 4. The resulting empirical models and statistical tools can be used to evaluate the test data and develop performance tolerance limits.

5.3   Measurement Quality Assurance

As discussed in Section 4.3, measurement quality assurance methods and procedures should be implemented to ensure that the testing program results in accurate measurements. This is achieved through the periodic calibration of the measuring equipment used for the final article

---

[13] Environmental testing is discussed in Section 4.2.2.

tests and the assessment of measurement uncertainty for the calibration results. Requirements for testing and calibration laboratories regarding measuring and test equipment calibration, measurement uncertainty and measurement traceability are contained in ISO/IEC 17025:2005 and ANSI/NCSLI Z540.3-2006. Periodic calibration and measurement traceability are also discussed in Section 10.4.

The data obtained from testing and calibration are not considered complete without statements of the associated measurement uncertainty. Therefore, test data cannot simply be used to establish tolerance limits for the test article attributes or parameters without an adequate assessment of measurement uncertainty. Similarly, calibration data alone cannot be used to verify conformance of test equipment to manufacturer specifications without evaluating the impact of measurement uncertainty on false accept or false reject risk.[14]

An in-depth coverage of key aspects of measurement uncertainty analysis and detailed procedures needed for developing such estimates is provided in NASA *Measurement Quality Assurance Handbook*, Annex 3 – *Measurement Uncertainty Analysis Principles and Methods*.

5.4   Specification Limits

In general, manufacturer specifications are intended to convey tolerance limits that are expected to contain a given performance parameter or attribute with some level of confidence under baseline conditions. These primary or baseline tolerance limits are typically established from functional tests conducted at nominal operating conditions. If so, they should be accompanied by a qualification statement indicating that all listed specifications are typical values referenced to standard conditions (e.g., 23 °C +/- 5 °C and 10 VDC excitation).

Additional modifier tolerance limits should be developed, as needed, to accommodate extended operating environments or to account for short-term mechanical stresses incurred during transport or use. For example, the modifier tolerance limits may correspond to temperature, shock and vibration parameters that affect the sensitivity and/or zero offset of a sensing device.

5.4.1   Probability Distributions

Performance parameters and attributes such as nonlinearity, repeatability, hysteresis, resolution, noise, thermal stability and zero shift are considered to be random variables that follow probability distributions that relate the frequency of occurrence of values to the values themselves. Therefore, the establishment of tolerance limits should be tied directly to the probability that a performance parameter or attribute will lie within these limits.

Probability distributions include, but are not limited to normal, lognormal, uniform (rectangular), triangular, quadratic, cosine, exponential and u-shaped. The selection of applicable probability distributions depends on the individual performance parameter or attribute and are often determined from test data obtained for a sample of articles or items selected from the production population. The sample statistics are used to infer information about the underlying parameter population distribution for the produced items. This population distribution represents the item to item variation of the given parameter.

---

[14] The application of uncertainty estimates for assessing the risk of falsely accepting or rejecting a calibrated parameter is discussed in Sections 10.4.5 and 11.3.

The performance parameter or attribute of an individual item may vary from the population mean. However, the majority of the produced items should have parameter mean values that are very close to the population mean. Accordingly, a central tendency exists that can be described by the normal distribution.

In cases, where asymmetry about the parameter population mean is observed or suspected, it still may be reasonable to assume that the normal distribution may be applicable. However, the lognormal or other asymmetric distribution may be more applicable.

There are a couple of exceptions when the uniform distribution would be applicable. These include digital output resolution and quantization resulting from the digital conversion of an analog signal.

5.4.2   Confidence Levels

Baseline performance specifications are often established from data obtained from the testing of a sample of items selected from the production population. Since the test results are applied to the entire population of produced items, the tolerance limits should be established to ensure that a large percentage of the items within the population will perform as specified. Consequently, the specifications are confidence limits with associated confidence levels.

Ideally, confidence levels should be commensurate with what the manufacturer considers to be the maximum allowable false accept risk (FAR). The general requirement is to minimize the probability of shipping an item with nonconforming (or out-of-compliance) performance parameters and attributes. In this regard, the primary factor in setting the maximum allowable FAR may be the costs associated with shipping nonconforming products.

For example, a manufacturer may require a maximum allowable FAR of 1% for all performance specifications. In this case, a 99% confidence level would be used to establish the parameter specification limits. Similarly, if the maximum allowable FAR is 5%, then a 95% confidence level should be used to establish the specification limits.

An in-depth coverage of the methods and principles of measurement decision risk analysis and the estimation and evaluation of FAR and false reject risk (FRR) are provided in NASA *Measurement Quality Assurance Handbook*, Annex 4 – *Estimation and Evaluation of Measurement Decision Risk*.

5.4.3   Confidence Limits

As previously discussed, performance parameter distributions are established by testing a selected sample of the production population. Since the test results are applied to the entire population of a given parameter, limits are developed to ensure that a large percentage of the population will perform as specified. Consequently, the parameter specifications are confidence limits with associated confidence levels.

As shown in Figure 5-1, the limits, $\pm L_x$, represent the confidence or containment limits for values of a specific performance parameter, x. The associated confidence level or containment probability[15] is the area under the distribution curve between these limits.

---

[15] In this context, confidence level and containment probability are synonymous, as are confidence limits and containment limits.

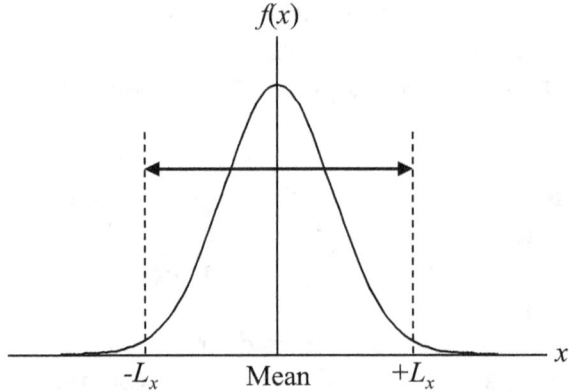

Figure 5-1 Normal Parameter Probability Distribution

For parameter populations that are normally distributed, the confidence limits can be established using the *t*-statistic.

$$\pm L_x = \pm t_{\alpha/2,v} \times s_x \qquad (5\text{-}1)$$

where

$t_{\alpha/2,v}$ = *t*-statistic
α = significance level = $1 - C/100$
C = confidence level (%)
v = degrees of freedom = $n - 1$
n = sample size
s_x = sample standard deviation

For parameter populations that are uniformly distributed, such as digital output resolution and quantization, the confidence limits would be established for a 100% containment probability, as shown in Figure 5-2.

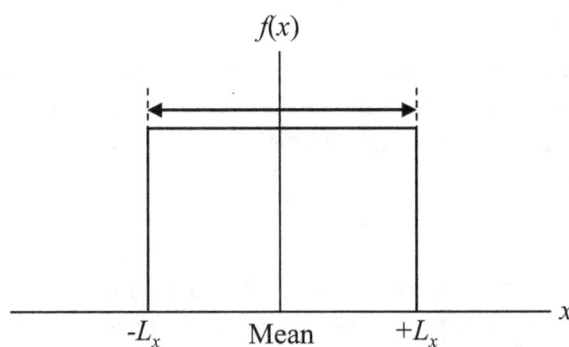

Figure 5-2 Uniform Parameter Probability Distribution

In this case, the confidence limits for $\pm L_{res}$ and $\pm L_{quan}$, would be established using the following equations.

$$\pm L_{res} = \pm \frac{h}{2} \qquad (5\text{-}2)$$

and

41

$$\pm L_{quan} = \pm \frac{A}{2^{n+1}} \qquad (5\text{-}3)$$

where

h = least significant display digit
A = full scale range of analog to digital converter
n = quantization significant bits.

5.4.3.1 Baseline Specification Limits

Baseline limits are typically established from data collected during functional testing conducted at nominal room temperature and relative humidity, with insignificant levels of electromagnetic interference, mechanical vibration and shock. Consequently, baseline limits are considered to be "best case" specification limits.

As previously discussed, test data are accompanied by measurement error. Therefore, baseline tolerance limits must also incorporate the uncertainty associated with measurement error. The preferred method is to combine the overall error distribution with the parameter distribution.

The total uncertainty for a measured parameter is comprised of uncertainties due to measurement equipment (e.g., bias and resolution error), repeatability or random error caused by fluctuations in environmental or other ancillary conditions, operator error, etc.

As shown in Figure 5-3, the combined probability distribution for three or more error sources begins to take on a Gaussian or normal shape, regardless of the shape of the individual error distributions. The combined error distribution takes on a non-Gaussian shape when one or more uniformly distributed errors are major contributors, but a central tendency still exists.[16] In either case, the total measurement uncertainty is equal to the standard deviation of the combined error distribution, σ_{ε_T}.

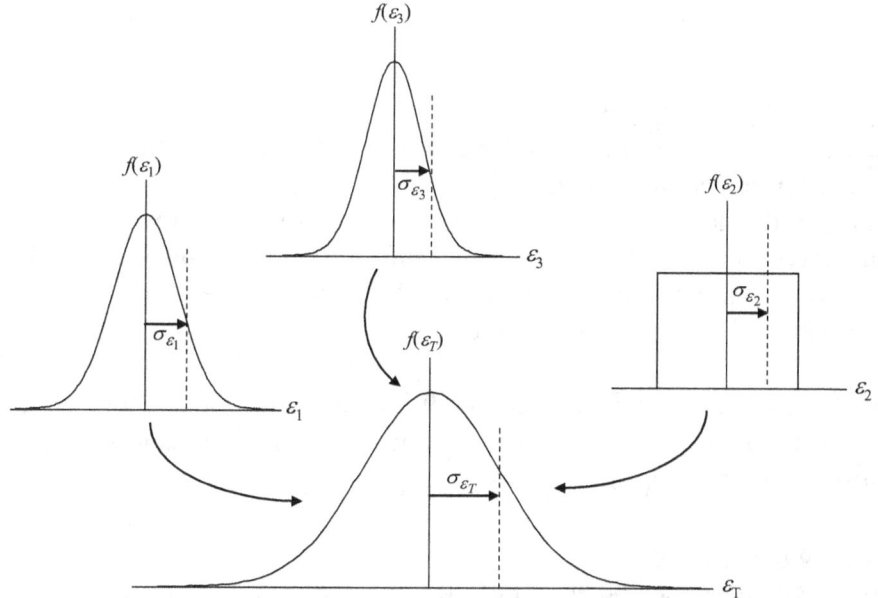

Figure 5-3 Combined Measurement Error Distribution

---

[16] See Castrup, H.: "Selecting and Applying Error Distributions in Uncertainty Analysis," presented at the Measurement Science Conference, Anaheim, CA, 2004.

Similarly, when the total error distribution and parameter distribution are combined, the resulting distribution will also exhibit a normal shape (albeit with a larger standard deviation), as shown in Figure 5-4.

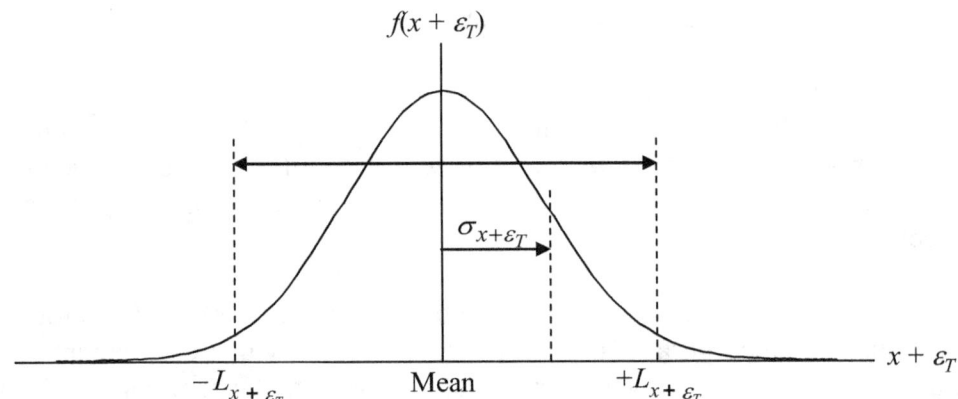

Figure 5-4 Combined Measurement Error and Parameter Distribution

The baseline limits are then computed from equation (5-4), using the standard deviation of the resulting distribution, the associated degrees of freedom and the specified confidence level.

$$\pm L_{x+\varepsilon_T} = \pm t_{\alpha/2,v} \times \sigma_{x+\varepsilon_T} \qquad (5\text{-}4)$$

where

$$
\begin{aligned}
t_{\alpha/2,v} &= t\text{-statistic} \\
\alpha &= \text{significance level} = 1 - C/100 \\
C &= \text{confidence level (\%)} \\
v &= \text{degrees of freedom} \\
\sigma_{x+\varepsilon_T} &= \text{distribution standard deviation}
\end{aligned}
$$

5.4.3.2 Secondary Specification Limits

Secondary tolerance limits account for possible drift or shift in performance parameters and attributes resulting from mechanical shock, vibration and other environmental conditions. These additional tolerance limits may be determined from test data to ensure that the product meets the overall performance requirements for a wider range of operating conditions. In some cases, however, these specification limits may be theoretically derived based on relevant data from components or materials used to develop the product.

The methods used to develop secondary tolerance limits are the same as those described for baseline limits. This includes the incorporation of uncertainty estimates to account for measurement errors encountered during testing.

5.4.3.3 Additional Modifier Limits

In some cases, modifier limits may be added to the baseline tolerance limits to account for the inherent variability in the manufacturing process. These additional limits, sometimes referred to as guardband limits, are ascertained during the production monitoring process. The development of modifier limits is discussed in Chapter 6.

CHAPTER 6: VERIFYING AND MODIFYING SPECIFICATIONS

The verification of MTE specifications is an important quality assurance requirement for both manufacturers and end users. Manufacturers conduct acceptance testing and monitoring processes to demonstrate that their products are free of defects and meet functional and performance specifications for a range of environmental and operating conditions.

Manufacturers also conduct engineering analyses to ensure that all relevant operating functions and performance specifications have been sufficiently evaluated before they are included in the final product documentation. End users conduct engineering analyses to assess whether or not the MTE performance specifications meet their application requirements.

Some MTE manufacturers will modify the baseline performance specifications to account for the inherent variability in the manufacturing process. These modifier limits are typically ascertained from production monitoring data. End users may modify MTE specification limits to reflect actual performance observed from calibration history data and to control measurement decision risk.

6.1   Acceptance Testing

Manufacturer acceptance testing includes performance demonstrations and environmental exposures to screen out manufacturing defects, workmanship errors, incipient failures and other performance anomalies that are not readily detected during normal inspections or basic functional tests. The acceptance testing conducted by MTE manufacturers prior to product shipment are similar to customer or end user acceptance testing.

As with final article testing discussed in Chapter 5, the test conditions are usually designed to be identical, or as close as possible, to the anticipated end usage environment, including extreme conditions. The objective is to ensure that the delivered product will meet or exceed the specified performance parameters. Acceptance testing provides a means of detecting product defects that might result from the manufacturing process and assessing how the product withstands a variety of shipping and storage conditions.

Acceptance testing for the evaluation of static, dynamic and environmental performance specifications is discussed in Chapter 10. Accelerated life testing and accelerated stress screening methods are discussed in Chapters 4 and 10.

6.2   Production Monitoring

The primary goal of production monitoring is to evaluate the manufacturing process through measurement and observation. The monitoring activities should cover the entire manufacturing process, including the assembly of previously tested and accepted components or subassemblies, to adequately assess the quality of the product.

The production monitoring program should verify functional and baseline performance specifications of the manufactured device or item. Some manufacturers may test the entire production population to ensure that individual items are performing within specified limits prior to shipment. In most cases, manufacturers will periodically test a randomly selected sample of the production population. The primary reasons for selected sampling include:

- The high cost of 100% product testing.

- The time required to test all products.

- The tests may intentionally or unintentionally induce product failures.

The sample size should be large enough to be representative of the entire production batch or lot and to provide sufficient test data for a meaningful statistical analysis. Production monitoring and data analysis methods can be found in various publications including the NIST *e-Handbook of Statistical Methods*.

The analysis results are used to identify nonconforming products and initiate immediate corrective action. If a unit fails during testing, then a failure mode analysis is conducted to establish the root cause(s). The batch of units, produced along with the failed unit, are then tagged for testing to verify the failure mode and repaired, if needed.

6.2.1 Sampling Procedures

A lot acceptance sampling plan (LASP) is developed to determine the proper disposition of the production lot (e.g., accept, reject, retest). The LASP uses a set of rules for making decisions based on counting the number of defects or non-conforming items in the sample.

The manufacturer establishes an acceptable quality level (AQL) that specifies the percentage of defective or non-conforming items that constitute the baseline quality requirement for the product. The manufacturer's sampling plan is designed so that there is a high probability of accepting a lot that has a defect level less than or equal to the AQL.

Conversely, the customer or consumer establishes a lot tolerance percent defects (LTPD) criterion so that there is a very low probability that a poor quality product is accepted. The LTPD is a designated high defect level that would be unacceptable to the customer. The customer would like the sampling plan to have a low probability of accepting a lot with a defect level as high as the LTPD.

The probability of wrongly accepting a defective lot (consumer's risk) or wrongly rejecting an acceptable lot (producer's risk) can be estimated for each sampling plan using the operating characteristics (OC) curve. The OC curve plots the probability of accepting the lot versus the percent of defects. The OC curve is the primary tool for displaying and investigating the properties of a LASP.

The producer would like to design a sampling plan such that the OC curve yields a high probability of accepting a good lot; while the customer wants to be protected from accepting poor quality products. MIL-STD-105E *Sampling Procedures and Tables for Inspection by Attributes*[17] has been used over the past few decades to achieve this goal for government procurement of materials and equipment.

In addition to MIL-STD-105E, it is common practice for manufacturers to use standardized sampling procedures such as the following:

- ANSI/ASQC Z1.4-2003 *Sampling Procedures and Tables for Inspection by Attributes*.

---

[17] Although MIL-STD-105E was officially cancelled in February 1995, many companies continue to use this standard for product inspection and testing.

- ANSI/ASQC Z1.9-2003 *Sampling Procedures and Tables for Inspection by Variables for Percent Nonconforming.*

- ISO 2859-1:1999 *Sampling procedures for inspection by attributes - Part 1: Sampling plans indexed by acceptable quality level (AQL) for lot-by-lot inspection.*

6.2.2   Statistical Quality Control

The purpose of statistical quality control is to ensure, in a cost effective manner, that the product shipped to the customer meets or exceeds its advertised performance specifications. Most of the statistical quality control methods used today were developed in the last century. Two common quality monitoring methods include statistical process control (SPC) charts and process capability indices.

In general, SPC charts looks at the sample mean values over time, while capability indices look and the sample mean and standard deviation of the sampled process. Both statistical quality control methods are discussed briefly in the following subsections.

6.2.2.1 <u>Statistical Process Control</u>

SPC refers to a number of different methods for monitoring manufacturing processes by establishing control limits that ensure that the key product parameters and attributes remain in-control. The product parameters are monitored using statistical control charts[18] to identify problems in a timely manner and to incorporate changes to the manufacturing process as needed.

The purpose of control charts is to monitor the uniformity of the process output, typically a product characteristic. Commonly used control charts include

- Shewhart $\bar{X}$ and s (mean and standard deviation)

- Shewhart $\bar{X}$ and R (mean and range)

- Exponentially weighted moving average (EWMA)

Statistical process control methods are detailed in various books and publications such as ASTM E2587-07 *Standard Practice for Use of Control Charts in Statistical Process Control.*

When monitoring manufactured MTE, performance parameters for sampled items are plotted along with the associated specification limits. The performance parameter is compared with the specification limits to see if it is in-control or out-of-control. Either single-parameter or multi-parameter control charts can be used to graphically monitor a statistic that represents more than one performance characteristic.

In general, control charts show the value of the performance parameter or attribute versus the sample number or versus time. As shown in Figure 6-1, a control chart contains a center line that represents the mean value for the in-control process, an upper control limit (UCL) line and a lower control limit line (LCL).

---

[18] The control chart was introduced in 1924 by Walter A. Shewhart while working at Bell Laboratories.

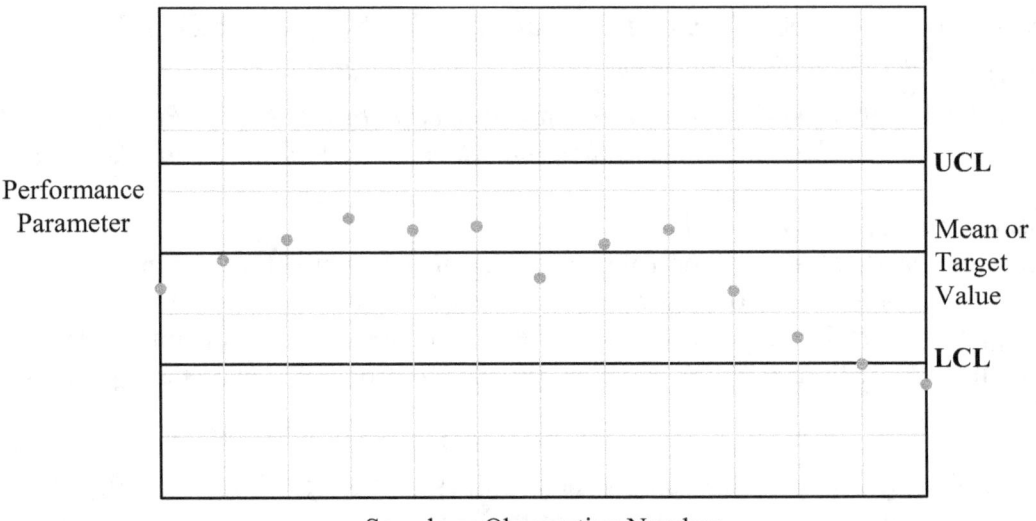

Performance Parameter

UCL

Mean or Target Value

LCL

Sample or Observation Number

Figure 6-1 General Control Chart Format

An in-control manufacturing process is one in which all of the test data (i.e., measurements) fall inside the specification limits. Using SPC methods alone, however, will not guarantee that a manufacturing process observed to be in-control won't produce a item or device that performs outside of its specification limits.

6.2.2.2 <u>Process Capability</u>
Once the production process has been observed to be in-control, the monitored performance parameter is compared to the specification limits to assess how well the process meets these limits. To do this, the distribution of the sampled parameter values and the specification limits are used to compute a capability index. The larger the capability index, the higher the probability that the manufactured product will meet or exceed its specified performance requirements.

There are different ways to compute the process capability index, but all methods assume that the underlying data population is normally distributed. Equations used to compute capability indices using process monitoring data are given below.

$$C_p = \frac{USL - LSL}{6s} \tag{6-1}$$

$$C_{pk} = \min \left| \frac{USL - \bar{x}}{3s}, \frac{\bar{x} - LSL}{3s} \right| \tag{6-2}$$

$$C_{pm} = \frac{USL - LSL}{6\sqrt{s^2 + (\bar{x} - T)^2}} \tag{6-3}$$

where

USL = Upper specification limit
LSL = Lower specification limit
$\bar{x}$ = Sample mean or average

47

s = Sample standard deviation
T = Target value

The C_p index relates the spread of the process or performance parameter values to the specification limits. In the calculation of C_p, it is assumed that the process mean is centered within the specification limits. The C_{pk} index accounts for both the process variation and the location of the process mean to the upper and lower specification limits. The C_{pm} index uses a modified process variation or standard deviation that accounts for differences between the process mean and target values.

Some industries require that manufacturers estimate and report capability indices, such as C_p or C_{pk}, to demonstrate that a statistically significant portion of their products will conform to specified performance limits. The estimation of process capability indices are detailed in various books and publications such as ASTM E2281-08 *Standard Practice for Process and Measurement Capability Indices*.

6.3   Measurement Quality Assurance

Companies and government organizations that maintain ISO 9001 or AS9100 quality management systems must maintain an effective testing program by confirming that the measuring equipment are in conformance with manufacturer specifications. This is achieved through

- The periodic calibration of test equipment to verify that they are capable of accurately and reliably performing their intended measurement tasks.

- The assessment of measurement uncertainty for all equipment calibrations using appropriate analysis methods and procedures that account for all sources of uncertainty (e.g., biases of reference standards or materials, display resolution, environmental conditions).

ISO/IEC 17025:2005 *General Requirements for the Competence of Testing and Calibration Laboratories* contains the requirements for testing and calibration laboratories must meet to demonstrate that they operate a quality program or system and are capable of producing technically valid results.

6.3.1   Test Equipment Calibration

ANSI/NCSLI Z540.3-2006 *Requirements for the Calibration of Measuring and Test Equipment* is the U.S. consensus standard that establishes the technical requirements for managing test equipment and assuring that their calibrated parameters conform to specified performance requirements.

ISO/IEC 17025 and ANSI/NCSLI Z540.3 also require that the measurement results obtained during equipment calibration must be traceable to a national measurement institute, such NIST. Calibration intervals or periods may vary for each device depending upon the stability, application and degree of use. Periodic calibration and measurement traceability are discussed in Section 10.2.

6.3.2 Measurement Uncertainty

Acceptance testing and production monitoring data cannot be properly analyzed without associated measurement uncertainty estimates. Similarly, test equipment calibration data alone cannot be used to verify conformance to manufacturer specifications without evaluating the impact of measurement uncertainty on false accept or false reject risk.[19]

An in-depth coverage of key aspects of measurement uncertainty analysis and detailed procedures needed for developing such estimates is provided in NASA *Measurement Quality Assurance Handbook*, Annex 3 – *Measurement Uncertainty Analysis Principles and Methods*.

6.4 Engineering Analysis

Baseline and secondary tolerance limits must be thoroughly analyzed and reviewed before the manufacturer publishes the MTE specifications. Similarly, potential users should conduct a technical assessment of the specifications to ensure that the MTE will meet the performance requirements for its intended application.

6.4.1 MTE Manufacturers

Manufacturers perform an engineering evaluation to determine if the MTE specification limits established during initial product development and testing sufficiently bound the majority of the acceptance testing and production monitoring data. The analysis should provide an impartial assessment of whether or not the required performance specifications are achieved in the final product.

The empirical models and statistical tools described in Chapters 4 and 5 may be used to analyze the data collected during acceptance testing or production monitoring and to modify MTE parameter specification limits as needed.

6.4.2 MTE End Users

End users must carefully review all of the specifications to assess whether the MTE will meet the measurement requirements of the intended application. When tolerance limits are specified as a percentage of full scale, then it is especially important to evaluate the specifications at the lower end of the operating range to ensure they meet the percentage of reading requirements.

Personnel involved in MTE selection should identify and combine all relevant specifications to determine if the accuracy requirements will be met during actual operating conditions. This analysis can also identify performance parameter that may need to be investigated through acceptance testing and calibration. Methods for combining MTE specifications are discussed in Chapters 9 and 11.

6.5 Modifying Specification Limits

As previously discussed, manufacturers may change one or more specification limits to better reflect observed performance during acceptance testing or production monitoring. For example, the actual spread in the parameter distribution may exceed the initial specification limits, as shown in Figure 6-2. In this case, the specification limits for performance parameter x would be increased to achieve the desired containment or in-tolerance probability.

---

[19] The application of uncertainty estimates for assessing the risk of falsely accepting or rejecting a calibrated parameter is discussed in Section 11.3.

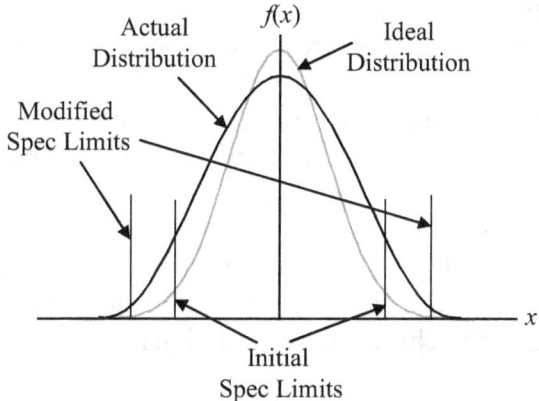

Figure 6-2 Modified Specification Limits

Any increase or decrease in specification limits must also account for the measurement uncertainty associated with the acceptance testing or production monitoring data to properly manage false accept or false reject risk.[20]

MTE users may modify the specification limits for one or more performance parameters to better reflect historical calibration data or the criticality of the measurement application. The increase or decrease in the specification limits are based on the maximum allowable false accept risk (FAR).

In this case, the confidence level should be commensurate with what the MTE user considers to be the maximum allowable FAR for the measurement application. Assuming an underlying normal distribution, the specification limits would be increased or decreased by the ratio of the *t*-statistics for the different confidence levels.

$$\pm L_{x,U} = \pm \frac{t_{\alpha_U/2,v}}{t_{\alpha_M/2,v}} \times L_{x,M} \qquad (6\text{-}4)$$

where

$L_{x,M}$ = Manufacturer specification limits for the MTE performance parameter x
$L_{x,U}$ = User modified specification limits
$t_{\alpha_M/2,v}$ = *t*-statistic corresponding to manufacturer confidence level, C_M
$t_{\alpha_U/2,v}$ = *t*-statistic corresponding to user confidence level, C_E
α_M = $1 - C_M$
α_U = user FAR = $1 - C_E$
v = Degrees of freedom used to establish $L_{x,M}$

If $C_U > C_M$, then $L_{x,U} > L_{x,M}$. Conversely, if $C_U < C_M$, then $L_{x,U} < L_{x,M}$. In either case, care must be taken to ensure that the modified specification limits meet the requirements of the measurement application. It is also important that the MTE user have a very good understanding of the manufacturer specifications and how they are established. Interpreting manufacturer specifications is discussed in Chapter 9.

---

[20] The application of uncertainty estimates for assessing the risk of falsely accepting or rejecting a calibrated or tested parameter is discussed in Sections 10.4.5 and 11.3.

6.6   Verification Reports

Reporting the results of acceptance testing, production monitoring or an engineering analysis is a major outcome of any product verification study. Verification reports should be comprehensive and clearly define the objectives of the study, describe the testing and/or analysis methods used to achieve these objectives, and concisely present the findings and results.

6.6.1   Acceptance Testing Report

The acceptance testing report should define the test objectives, identify and describe the configuration of the item(s) tested, document the test plan, and specify the environmental test conditions and procedures used to achieve these objectives. The acceptance testing report should include the following:

- An executive summary that highlights the acceptance testing program, the experimental design approach, data analysis methods used and the main findings and conclusions.

- An introduction explaining the purpose, objectives and scope of the test program.

- A description of the MTE parameters tested, the environmental conditions and duration of the tests, and the acceptance or rejection criteria used.

- A description of the experimental design approach and supporting diagrams, schematics or drawings depicting the test article(s), the overall test setup and the specific test equipment or reference standards used.

- A discussion of the data analysis methods used, along with any underlying assumptions or simplifications applied, and a summary of the data analysis results highlighting expected and unexpected findings, supported by graphs, tables and charts, and a comparison to other related experimental studies or engineering analyses, if applicable.

- Additional discussions identifying and describing any detected deficiencies, failures or unsatisfactory performance encountered during testing, along with recommendations for corrective actions or the brief synopsis of any corrective actions that have been implemented.

- Conclusions summarizing the overall findings, including the recommended or modified tolerance limits within which each tested parameter or characteristic is considered to provide acceptable performance, and any recommendations for additional testing and/or analysis, if needed.

6.6.2   Production Monitoring Report

The production monitoring report should define the objectives of the monitoring program, identify and describe the configuration of the item tested, and specify the methods and procedures used to achieve these objectives. The report should document whether the manufactured product will or will not satisfy the specified performance requirements. The production monitoring report should include the following:

- An executive summary that highlights the main objectives of the monitoring program, the MTE parameters tested, the data analysis methods used and the main findings and conclusions.

- A description of the MTE parameters tested, the environmental conditions and duration of the tests, and the testing criteria used.

- A description of the test equipment or reference standards used.

- A discussion of the data analysis methods, along with any underlying assumptions or simplifications applied, and a summary of the data analysis results highlighting expected and unexpected findings, supported by graphs, tables and charts, and a comparison to other related experimental studies or engineering or analyses, if applicable.

- Additional discussions identifying and describing any detected deficiencies, failures or unsatisfactory performance encountered during testing, along with recommendations for corrective actions or a synopsis of any corrective actions that have been implemented.

- Conclusions summarizing the overall findings, including the recommended or modified tolerance limits within which each MTE parameter or characteristic is considered to provide acceptable performance.

6.6.3   Engineering Analysis Report
The contents of the engineering analysis report should include:

- An executive summary that highlights the main objectives of the analysis, the methodology used and the main findings and conclusions.

- An introduction explaining the purpose and scope of the study.

- An assessment of the MTE specification documents, user manuals, operating manuals, drawings and other related technical information.

- A description of empirical models, statistical tools or other analysis methods used.

- A discussion of the analysis results highlighting expected and unexpected findings, supported by graphs, tables and charts, and a comparison to related product tests or other engineering studies, if applicable.

- Conclusions summarizing the overall findings, including the recommended or modified tolerance limits within which each MTE parameter is considered to provide acceptable performance, and any recommendations for testing and/or additional analysis, if needed.

- Appendices containing raw data, sample calculations, and other supplemental information.

CHAPTER 7: REPORTING SPECIFICATIONS

The preparation and dissemination of comprehensive specification documents are an important element of the MTE design, manufacturing and marketing processes. Manufacturer/model specifications should be reported in a logical format, using consistent terms, abbreviations and units that clearly convey pertinent performance parameters and attributes.

There are differences of opinion regarding which MTE parameters and attributes should be specified and how they should be reported. This is evident from the inconsistency in the manufacturer specifications for similar equipment.[21] Regardless, specification documents should contain all relevant aspects and information required to evaluate the suitability of the MTE for the intended application.

Given the complexity of present day MTE there is a need for standardized specification formats. Technical organizations, such as ISA and SMA, have published documents that adopt standardized instrumentation terms and definitions. ISA has also developed standards that provide uniform requirements for specifying design and performance characteristics for selected electronic transducers.[22] Similarly, SMA has developed a standard for the data and specifications that must be provided by load cell manufacturers.

Despite these few exceptions for electronic transducers, the vast majority of specification documents fall short of providing the information required to evaluate the suitability of MTE for a given application. For example, it is common for manufacturers to omit information about the underlying probability distributions for the MTE performance parameters. Additionally, manufacturers do not often report the corresponding confidence level for the specified MTE parameter tolerance limits.

7.1   Operating Principles

The scientific basis or fundamental mechanism(s) upon which the MTE operates or functions should be conveyed in an unambiguous manner. This may include descriptions of the sensing element(s) and other key components, as well as a mathematical equation or transfer function that relates the input and output of the measuring device.

7.1.1   Physical Design Characteristics

The theory, methods, concepts and materials employed in the MTE should be described, along with other physical, electrical and/or dimensional design characteristics. This information should provide sufficient detail to help in the MTE selection process.

For example, a load cell might be described as a strain-gage based force transducer that requires an external excitation voltage. The basic design information should include the load cell type (e.g., tension, compression or universal), the type of strain-gage used (e.g., metallic or semiconductor, bonded or un-bonded, wire or foil) and the number of active strain-gage bridge arms.

---

[21] Similar equipment constitutes MTE from different manufacturers that can be substituted or interchanged without degradation of measurement capability and quality.

[22] A list of ISA transducer standards is given in Appendix B.

7.1.2 Transfer Function

The theory and principle of operation should also be expressed as a transfer function that describes the mathematical relationship between the input and output response of the device under ideal or stated operating conditions. For example, the load cell transfer function shown in equation (7-1) relates the input load to the output voltage.

$$LC_{out} = W \times S \times V_{Ex} \tag{7-1}$$

where

LC_{out} = Output voltage
W = Applied load or weight
S = Load cell sensitivity
V_{Ex} = Excitation voltage

In this case, is also important to include a note indicating that the validity of the load cell transfer function depends on the use of appropriate units for the variables, W, S and V_{Ex}. If the load cell has a rated output expressed in mV/V for loads expressed in lb_f, then the load cell sensitivity should be expressed in $mV/V/lb_f$.

7.2 Performance Characteristics

Both the static and dynamic performance characteristics of the MTE should be specified. The static performance characteristics (e.g., sensitivity, zero offset) should provide an indication of how the equipment or device responds to a steady-state input at one particular time. The dynamic performance characteristics (e.g., response time, zero drift) should provide an indication of how the equipment or device responds to changes in input over time.

7.2.1 Baseline Specifications

The baseline specifications should consist of the parameters, and associated tolerance limits, that affect the basic MTE functionality and performance under nominal or typical operating conditions. Of course, what comprises typical operating conditions must also be specified (e.g., 23 °C ± 5 °C and maximum 70% relative humidity).

Static baseline specifications include, but are not limited to

- Sensitivity or Gain
- Zero Offset
- Nonlinearity
- Hysteresis
- Repeatability or Noise
- Resolution

Dynamic baseline specifications include, but are not limited to

- Response Time
- Frequency Response
- Cutoff Frequencies
- Passband Ripple
- Damping
- Phase Shift

Each MTE is designed to make measurements over a certain range of input values. The baseline specifications should include the upper and lower limits of the input range of the device. For example, a piezoelectric transducer may be designed to measure pressures from 0 to 100 psig. Operating the transducer beyond these limits may damage the transducer or permanently change one or more of the baseline performance characteristics. If there is an overpressure range that can be tolerated before permanent damage occurs, then it should also be stated.

The output range of the measuring device, which is proportional to the input range, should also be stated. For example, the output range for the pressure transducer may be 0 to 100 millivolts (mV). In this case, the full scale output would be specified as 100 mV.

7.2.2 Secondary Specifications

Secondary or modifier specifications should be included to account for the drift or shift in baseline performance characteristics resulting from mechanical shock, vibration, temperature, humidity and other environmental conditions. These secondary specification limits are combined with the baseline specification limits when the MTE is used outside the associated baseline conditions.

Secondary or modifier specifications include, but are not limited to

- Thermal Sensitivity Shift
- Thermal Zero Shift
- Acceleration Sensitivity
- Vibration Sensitivity

If the MTE incorporates hardware and/or software to compensate or "correct" for the effects of temperature on performance, then the compensated temperature range must be specified. And, since compensation methods cannot entirely eliminate error due to temperature effects, a temperature correction error (or temperature coefficient) must also be specified.

7.2.3 Time-dependent Specifications

Short-term and long-term drift or shift in MTE performance caused by usage conditions and component aging should be specified. Stability specifications provide an indication of the ability of the MTE to maintain its performance characteristics over an extended period of time.

The specification document should include a time period during which the MTE performance can be expected to conform to the stated tolerance limits. The time period provided should account for the drift rate inherent to the device or instrument. In some cases, it may be necessary to specify the performance parameters for several time periods (e.g., 30, 90, 180 and 360 days).

As previously discussed, the sensitivity and/or zero offset of a device or instrument can also change or drift due to environmental operating conditions. The influence of time-dependent and environmental influences are often interrelated, which may make it difficult to specify them separately.

7.3 Operating Conditions

The range of environmental conditions to which the MTE can be subjected, without permanently affecting performance or causing physical damage, must be specified. The specified environmental and other operating conditions should include, but are not limited to

- Operating Temperature Range
- Operating Pressure Range
- Operating Humidity Range
- Storage Temperature Range
- Maximum Shock or Vibration
- Maximum Acceleration

External power or warm-up requirements that affect MTE performance should also be specified, if applicable. For example, some transducers may require an external power supply such as voltage or current.

Excitation voltage specifications should include the voltage level, accuracy and maximum current. Excitation current specifications should include the current level, accuracy and the maximum load resistance. The excitation specification should also include any voltage or current drift rate with temperature, if applicable.

7.4   Terms, Definitions and Abbreviations

Specification documents should include all necessary terms, definitions and abbreviations to facilitate the interpretation and application of the stated performance characteristics and associated information. When possible, the terms, definitions and abbreviations should be consistent with standard practices.

Technical organizations, such as ISA and SMA, have published documents that adopt standardized instrumentation terms and definitions.[23],[24] However, there may be a need for further clarification and consistency in the general terms and definitions used in the reporting of MTE specifications. For example, if the terms *typical* or *nominal* are used to indicate the expected performance of the MTE parameters or attributes, then they should be clearly defined.

MTE specifications often include the use of acronyms or abbreviations such as FS (full scale), RDG (reading), RO (rated output), and BFSL (best-fit straight line). It is important that abbreviations be clearly defined in the specification document.

7.5   Probability Distributions

MTE performance parameters such as nonlinearity, repeatability, hysteresis, resolution, noise, thermal stability and zero shift are considered to be random variables that follow probability distributions that relate the frequency of occurrence of values to the values themselves. Therefore, MTE specifications should state the underlying probability distribution that was used to establishment tolerance limits for each performance parameter.

These distributions are necessary for proper interpretation and application of MTE specifications. In particular, probability distributions must be known to estimate MTE parameter bias and resolution uncertainties, to correctly combine parameter tolerance limits, to compute test

---

[23] ISA-37.1-1975 (R1982): *Electrical Transducer Nomenclature and Terminology*, The Instrumentation, Systems and Automation Society.

[24] SMA LCS 04-99: Standard Load Cell Specification, Scale Manufacturers Association.

uncertainty ratios and to estimate the risk of falsely accepting a performance parameter as being in-tolerance during calibration.

7.6   Confidence Levels

As discussed in Chapter 5, MTE specifications are tolerance limits established to ensure that a large percentage of the parameter population will meet performance requirements. Consequently, the specifications are confidence limits with associated confidence levels. The confidence level for a specified MTE parameter should be commensurate with what the manufacturer considers to be the maximum allowable false accept risk (FAR).

For example, a manufacturer may require a maximum allowable FAR of 1% for all performance specifications. In this case, the specification document should state that a 99% confidence level was used to establish the parameter tolerance limits. Similarly, if the maximum allowable FAR is 5%, then the specification document should state that a 95% confidence level was used to establish the parameter tolerance limits.

7.7   Specification Units and Conversion Factors

As with terms and definitions, specification units for all performance parameters and attributes should be consistent and clearly stated to facilitate their interpretation and application. Unit conversion factors should also be included in cases where they are needed to estimate MTE bias uncertainty or combine tolerance limits.

MTE performance parameters such as nonlinearity, hysteresis and repeatability are often specified as a percentage of full scale output because it results in a lower value than if specified as a percentage of reading. In this case, it is important to clearly state whether the specification is a percentage of full scale (% FS), reading (% RDG), rated output (% RO) or other operating range.

Noise specifications such as normal mode rejection ratio (NMRR) and common mode rejection ratio (CMRR) are generally specified in decibels (dB) at specified frequencies (usually 50 and 60 Hz). However, NMRR and CMRR should be specified for the entire measurement frequency range. The dB is a dimensionless logarithmic unit used to describe ratios of power, voltage or sound pressure.

dB_m = 10 log(P/P_0) where P_0 is the reference power level of 1 mW.

dB_V = 10 log (V^2/V_0^2) = 20 log (V/V_0) where V_0 is the reference voltage level.

dB_A = 10 log (p^2/p_0^2) = 20 log (p/p_0) where p_0 is the reference sound pressure of 20 micropascals (μPa).

The specification document should include the mathematical relationship between the dB unit and the associated voltage, sound or power unit as shown below.

$$P = 10^{\frac{dB_m}{10}} \times P_0 \qquad\qquad (7\text{-}1)$$

$$V = 10^{\frac{dB_V}{20}} \times V_0 \qquad\qquad (7\text{-}2)$$

57

$$p = 10^{\frac{dB_A}{10}} \times p_0 \tag{7-3}$$

where

P_0 = 1 mW
V_0 = Reference voltage level
p_0 = 20 μPa

If the numerical value of the reference is not provided, as in the dB gain of an amplifier, then it should be stated that the decibel specification is purely relative. In this case, equation (7-4) applies.

$$G = 10^{\frac{dB}{10}} \tag{7-4}$$

7.8   Specification Format

As previously stated, there is no universal guide or standard regarding the content or format of MTE specification documents. The benefits of a standardized specification format include

- The uniform presentation of descriptive and technical information.
- The use of uniform terminology, symbols and units.
- The direct comparison of similar equipment from different manufacturers.
- A systematic, comprehensive approach to the preparation of specification documents.

While various specification formats currently exist, there are some practical guidelines that should be followed.

1. The first page should include the basic design information, including product pictures or schematics.

2. The following page(s) should provide a tabulated summary of the performance parameter specifications and operating conditions. The applicable confidence level should be included in the specification table.

3. All footnotes, caveats, qualifications and stipulations regarding the published performance parameters should immediately follow the tabulated specifications. The probability distribution(s) applicable to the specified parameters should also be included in the footnotes.

4. Additional pages describing the operating principles and functional characteristics of the device should be included as needed.

5. The final page should include terms, definitions, abbreviations and unit conversions as needed.

CHAPTER 8: OBTAINING SPECIFICATIONS

It is important to obtain all relevant performance specifications prior to MTE selection or use. MTE specifications can be obtained from a variety of sources, including manufacturer websites, sales literature, catalogs and operating manuals. Manufacturers will also provide MTE specifications upon request by phone, fax or email.

MTE specification data sheets, operating manuals and other related technical documents are often given a document control number and issuance date as part of the manufacturer's configuration management program. When obtaining specifications, care must be given to ensuring configuration control between the published document(s) and the MTE model number, serial number, functions and options.

8.1   Product Data Sheets

Product data sheets comprise one to two pages summarizing MTE features, characteristics, specifications and environmental operating conditions. An example of a load cell specification data sheet is shown in Figure 8-1.

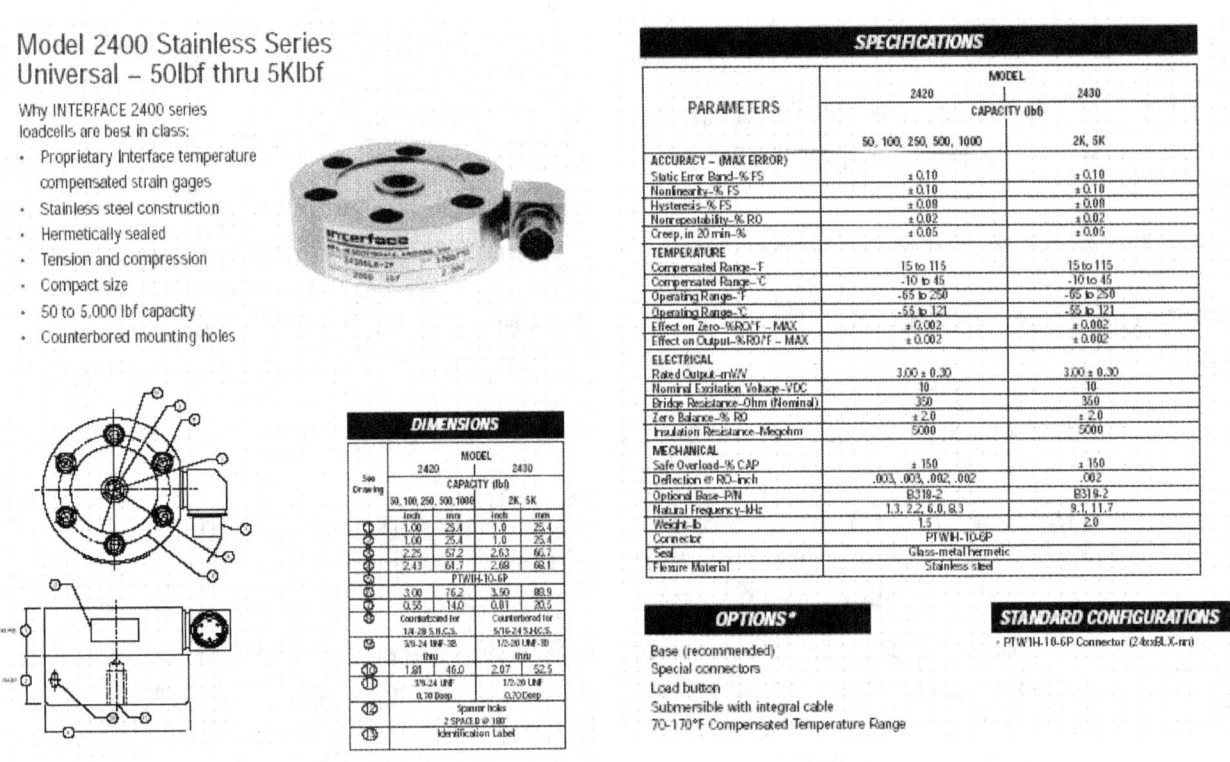

Figure 8-1 Load Cell Product Data Sheet

Most manufacturers publish MTE specifications on their web pages in Adobe Acrobat portable document format (PDF) files that can be directly downloaded. While a step up from advertising literature, product data sheets are often developed (or heavily influenced) by the marketing and sales department.

Therefore, product data sheets should not be used as the primary source for MTE performance specifications. It is also important to read all footnotes on the product data sheet, especially if they pertain to the interpretation or application of the specifications. Footnotes also often include caveats and warnings that the specifications may be subject to change without notice.

8.2   User Guides and Manuals

Equipment manufacturers often provide user guides and manuals that contain more complete descriptions of features and functions, as well as the performance specifications. These product guides and manuals also often include schematics and figures, step-by-step procedures and other technical information to facilitate proper MTE operation and maintenance.

MTE user guides and manuals are written by engineers familiar with the operating principles, functionality and performance requirements of the equipment. Manufacturers generally agree that the specifications and operating characteristics published in these documents supersede the information contained in product data sheets or other promotional literature.

A variety of equipment user guides and manuals are available in PDF format that can be downloaded from manufacturer websites. This makes it easy to obtain the technical details needed to evaluate equipment for the required measurement application.

8.3   Application Notes

Manufacturers may provide application or product notes that are typically one or two pages describing specific techniques or methods for using the product. Some application notes provide troubleshooting tips or address common maintenance issues that users may have encountered.

Application notes generally provide supplemental information to specification data sheets and user manuals. They are often available as PDF files for download from manufacturer websites. However, most manufacturers develop application notes to promote key characteristics and features of their product. Therefore, they are considered to be more akin to marketing literature.

8.4   Technical Notes and Articles

Technical notes and articles are often available as PDF files for download from manufacturer websites. Technical notes are typically similar in content and purpose to application notes, while technical articles and papers tend to provide more useful information about the theory and fundamental operating principles of the equipment or device.

Technical articles and papers may also include design details describing internal materials, parts and components. The most useful articles or papers are those that report actual equipment testing and evaluation. While a technical assessment provided by an equipment manufacturer may not be unbiased or independent, it can provide additional insight into important operational aspects not found elsewhere.

8.5   Other Resources

Obtaining specifications and user manuals for older equipment can be a challenging task. Some manufacturers maintain an archive of downloadable specification data sheets and user manuals for obsolete or discontinued products. Most manufacturers can provide specifications and related information for discontinued products upon request by phone, fax or email.

8.5.1   Third Party Websites

Specifications and product information for test and measurement equipment can also be accessed from other searchable websites such as www.thomasnet.com, www.thomasglobal.com and www.globalspec.com.

The ThomasNet and ThomasGlobal websites are based on the *Thomas Register of American Manufacturers*, a multi-volume directory of industrial product information covering 650,000 distributors, manufacturers and service companies within 67,000-plus industrial categories. GlobalSpec offers a similar web-based searchable database. These online information services are free of charge, but require registration before they can be utilized.

8.5.2 Engineering Handbooks

The physical, electrical, thermal and chemical properties of fundamental substances and materials can be obtained from engineering handbooks published by various professional organizations and government agencies. For example, NIST Monograph 180, *The Gauge Block Handbook*, and ASME B89.1.9 – 2002, *Gage Blocks*, specify tolerance limits for the length, flatness, parallelism and other characteristics of different gage block grades and sizes.

Selected handbook sources are listed below.

- National Institute of Standards and Technology, Gaithersburg, MD, www.nist.gov
- American Society of Mechanical Engineers, New York, NY. www.asme.org
- Institute of Electrical and Electronics Engineers, Piscataway, NJ, www.ieee.org
- American Institute of Chemical Engineers, New York, NY, www.aiche.org
- International Society for Instrumentation, Systems, and Automation, Research Triangle Park, NC, www.isa.org

CHAPTER 9: INTERPRETING SPECIFICATIONS

Manufacturer specifications are used to purchase MTE or to select suitable substitute equipment for a given measurement application, estimate bias uncertainties and establish tolerance limits for calibration and testing. When designing measurement systems, specifications provide a means of predicting overall performance and anticipating possible component problems.

Therefore, MTE users must be proficient at identifying applicable specifications and in interpreting and combining them. A basic understanding of the fundamental operating principles of the MTE is an important requirement for the proper interpretation of performance specifications. It is also important that manufacturers and users have a good understanding and assessment of the confidence levels and error distributions applicable to MTE specifications.

In select instances, the information included in a specification document may follow a recommended format.[25] However, the vast majority of specification documents fall short of providing crucial information about the confidence levels associated with reported specification limits. MTE manufacturers also do not indicate the applicable probability distribution for specified performance parameters. The lack of a universal standard regarding the development and reporting of MTE specifications has also resulted in inconsistent terms and units.

Consequently, it is often difficult to interpret and apply MTE specifications without gaining further clarification or making some underlying assumptions. It is always a good practice to

1. Review the specifications and highlight the performance parameters that need clarification.

2. Check the operating manual and associated technical documents for other useful details.

3. Request additional information and clarification from the manufacturer's technical department.

Ultimately, MTE users must determine which specifications are relevant to their application.

9.1   Operating Principles

A fundamental understanding of the measurement theory, concepts and components employed provides an important means of evaluating MTE performance specifications. In some cases, first-hand experience about the equipment or device may be gained through calibration and testing. In other cases, detailed knowledge about the MTE may be obtained from operating manuals, training courses, patents and other technical documents provided by the manufacturer.

Product data sheets typically provide a basic description of the fundamental mechanism(s) by which the measuring device operates. Information about the physical design characteristics, sensing elements, components and materials employed in the measurement device can often be found in MTE operating manuals. Application notes and technical articles published by the equipment manufacturer may also include design details describing internal materials, parts and components.

Technical standards, books and journal articles can provide useful information about the theory

---

[25] See for example, ISA-RP37.2-1982-(R1995): *Specifications and Tests for Strain Gauge Pressure Transducers.*

and fundamental operating principals of a measuring device. Journal articles and papers written by equipment users, and focused on equipment testing and evaluation, can provide additional insight into important operational aspects of the device.

Product patents are an additional, often overlooked technical resource. The United States Patent and Trademark Office website (www.uspto.gov) provides a searchable database of nearly eight million patents. Similarly, Google has a patent search feature (www.google.com/patents) that is very easy to use.

Information about the theory and principle of operation should then be converted into a transfer function that describes the mathematical relationship between the input and output response of the measuring device. This equation is the basis for interpreting, combining and applying MTE specifications.

The form of the transfer function depends on the type of MTE. As discussed in Chapter 7, a strain-gage based load cell has the basic transfer function

$$LC_{out} = W \times S \times V_{Ex} \tag{9-1}$$

where

$$
\begin{aligned}
LC_{out} &= \text{Output voltage} \\
W &= \text{Applied load or weight} \\
S &= \text{Load cell sensitivity} \\
V_{Ex} &= \text{Excitation voltage}
\end{aligned}
$$

A list of generalized transfer functions is provided in Appendix C for illustrative purposes. These equations are not considered to be definitive because the actual form of the transfer function will vary depending on the MTE specifications. In Section 9.2, the load cell transfer function is modified to account for performance specifications.

9.2   Performance Characteristics

For the most part, manufacturer specification limits are expected to contain the performance characteristics of the MTE parameters or attributes with some level of confidence. For example, these limits may correspond to temperature, shock and vibration parameters that affect the sensitivity and/or zero offset of a sensing device.

The load cell transfer function given in equation (9-1) indicates that the applied load, sensitivity and excitation voltage impact the output response of the load cell. Given the performance specifications[26] listed in Table 9-1, the load cell has a rated output of 2 mV/V for loads up to 5 lb$_f$ which equates to a nominal sensitivity of 0.4 mV/V/lb$_f$.

According to the specifications, the load cell output will also be affected by the following performance parameters:

- Excitation Voltage
- Nonlinearity
- Hysteresis

---

[26] Specifications obtained from www.ttloadcells.com/mdb-load-cell.cfm

- Noise
- Zero Balance
- Temperature Effect on Output
- Temperature Effect on Zero

Table 9-1. Specifications for MDB-5-T Load Cell

Specification	Value	Units
Rated Output (R.O.)	2 (nominal)	mV/V
Maximum Load	5	lb$_f$
Nonlinearity	0.05% of R.O.	mV/V
Hysteresis	0.05% of R.O.	mV/V
Nonrepeatability (Noise)	0.05% of R.O.	mV/V
Zero Balance (Zero Offset)	1.0% of R.O.	mV/V
Compensated Temp. Range	60 to 160	°F
Safe Temperature Range	-60 to 200	°F
Temperature Effect on Output	0.005% of Load/°F	lb$_f$/°F
Temperature Effect on Zero	0.005% of R.O./°F	mV/V/°F
Required Excitation	10	VDC

Equation (9-1) must be modified to account for these performance parameters. Given the assortment of specification units, it is apparent that the parameters cannot simply be added at the end of equation (9-1). The appropriate load cell output equation is expressed in equation (9-2).

$$LC_{out} = \left[\left(W + TE_{out} \times TR_{°F} \right) \times S + NL + Hys + NS + ZO + TE_{zero} \times TR_{°F} \right] \times V_{Ex} \qquad (9\text{-}2)$$

where

LC_{out} = Output voltage, mV
W = Applied weight or load, lb$_f$
TE_{out} = Temperature effect on output, lb$_f$/°F
$TR_{°F}$ = Temperature range, °F
S = Load cell sensitivity, mV/V/lb$_f$
NL = Nonlinearity, mV/V
Hys = Hysteresis, mV/V
NS = Noise and ripple, mV/V
ZO = Zero offset, mV/V
TE_{zero} = Temperature effect on zero, mV/V/°F
V_{Ex} = Excitation voltage, V

Equation (9-2) constitutes the modified transfer function for the load cell output. Given some knowledge about the probability distributions for the performance parameters, the uncertainty in the load cell output voltage for a given applied load can be estimated.[27] The uncertainty analysis procedure for the load cell is illustrated in Chapter 11. Probability distributions are discussed further in Section 9.8.

---

[27] If the load cell is tested or calibrated using a weight standard, then any error associated with the weight should also included in the modified transfer function.

9.3 Operating Conditions & Related Specifications

As seen from equation (9-2), environmental and other operating conditions can impact the output of a measuring device. Therefore, it is important to determine the actual or expected ranges of environmental conditions in which the MTE will be used. These ranges will determine the impact of secondary performance parameters such as thermal sensitivity shift and thermal zero shift.

The external power supply and related specifications should also be carefully evaluated, if applicable. Additional specifications, such as warm-up period, response time and the maximum allowable input can also affect the performance and safe operation of the device.

9.4 Terms, Definitions and Abbreviations

Technical organizations, such as ISA and SMA, have published documents that adopt standardized instrumentation terms and definitions.[28,29] However, there is a need for further clarification and consistency in the general terms and definitions used in the reporting of MTE specifications. There are particular terms and abbreviations that require further discussion.

For example, some MTE specifications may convey performance characteristics as "typical" or "maximum" values. However, the basis for these classifications is not often apparent and introduces confusion about which specification (typical or maximum) is applicable. In addition, the associated confidence levels, containment probabilities or coverage factors are often not provided, making it difficult to interpret either set of specifications. Consequently, manufacturers must be contacted for clarification.

MTE specifications commonly include the use of abbreviations such as FS, FSO, FSI, RDG, RO, RC and BFSL. The abbreviation FS (or F.S.) refers to full scale. Similarly, the abbreviation FSO (or F.S.O.) refers to full scale output and the abbreviation FSI (or F.S.I.) refers to full scale input. Specifications that are reported as % FS (or ppm FS) generally refer to full scale output. When in doubt, however, contact the manufacturer for clarification.

The abbreviation RDG refers to reading or output value. The abbreviation RO (or R.O.) refers to rated output and the abbreviation RC (or R.C.) refers to rated capacity. In some cases, the abbreviation BFSL is used to indicate that a combined non-linearity, hysteresis and repeatability specification is based on observed deviations from a best-fit straight line. Abbreviations commonly used in MTE specification documents are listed in Table 9-2.

Table 9-2. Common MTE Abbreviations

Abbreviation	Description	Abbreviation	Description
ADC	Analog to Digital Converter	NMV	Normal-Mode Voltage
BFSL	Best-fit Straight Line	ppb	Parts per billion
CMR	Common-Mode Rejection	ppm	Parts per million
CMRR	Common-Mode Rejection Ratio	RDG	Reading
CMV	Common-Mode Voltage	RF	Radio Frequency

---

[28] ISA-37.1-1975 (R1982): *Electrical Transducer Nomenclature and Terminology*, The Instrumentation, Systems and Automation Society.

[29] SMA LCS 04-99: *Standard Load Cell Specification*, Scale Manufacturers Association.

Abbreviation	Description	Abbreviation	Description
CTE	Coefficient of Thermal Expansion	RH	Relative Humidity
DAC	Digital to analog converter	RO	Rated Output
FS	Full Scale	RSS	1) Root-sum-square 2) Residual sum of squares
FSI	Full Scale Input	RTI	Referred to Input
FSO	Full Scale Output	RTO	Referred to Output
FSR	Full Scale Range	TRS	Transverse Rupture Strength
LSB	Least Significant Bit	VAC	Alternating Current Volts
LSD	Least Significant Digit	VDC	Direct Current Volts
NMRR	Normal-Mode Rejection Ratio		

9.5   Specification Units

As with terms and definitions, specification units can vary between manufacturers of similar MTE models. In addition, specification units can vary from one performance parameter to another for a given manufacturer/model.

For example, display resolution specifications can be expressed in digits, counts, percent (%) or other units such as mV or °C. Nonlinearity, hysteresis and repeatability specifications can be expressed as % FS, ppm FS, % RDG, ppm RDG, % RO or other units. Sensitivity specifications can be expressed as mV/psi, mV/g, μV/°F, pC/g or other units.

Specifications related to environmental conditions can be expressed as % FS/°F, % RO/°C, ppm/°C, % FS/g, psi/g, psi/°F, mV/°C, %Load/°F, etc. Noise specifications such as NMRR and CMRR are generally specified in decibels (dB) at specified frequencies (usually 50 and 60 Hz).

The use of different units, while often necessary, can make it especially difficult to interpret specifications. In most cases, units conversion is required before specifications can be properly applied.[30] Selected specification conversion factors are listed in Table 9-3 for illustration.

Table 9-3. Specification Conversion Factors

Percent	ppm	dB	Relative to 10 V	Relative to 100 psi	Relative to 10 kg/°C
1%	10000	-40	100 mV	1 psi	100 g/°C
0.1%	1000	-60	10 mV	0.1 psi	10 g/°C
0.01%	100	-80	1 mV	0.01 psi	1 g/°C
0.001%	10	-100	100 μV	0.001 psi	100 mg/°C
0.0001%	1	-120	10 μV	0.0001 psi	10 mg/°C

The dimensionless dB unit in Table 9-3 expresses the ratio of two values of power, P_1 and P_2, where dB = $10 \log(P_2/P_1)$. The dB values are computed for P_2/P_1 ratios corresponding to the

---

[30] Additional calculations may be required before specifications can be properly used to estimate parameter bias uncertainties and tolerance limits.

percent and ppm values listed. For electrical power, it is important to note that power is proportional to the square of voltage, V, so that $dB_V = 10 \log (V_1^2/V_2^2) = 20 \log (V_1/V_2)$. Similarly, acoustical power is proportional to the square of sound pressure, p, so that $dB_A = 10 \log (p_1^2/p_2^2) = 20 \log (p_1/p_2)$.

9.6   Qualifications, Stipulations and Warnings

Most MTE specifications describe the performance parameters and attributes covered by the manufacturer's product warranty. These reported specifications also often include qualifications, clarifications and/or caveats. Therefore, it is a good practice to read all notes and footnotes carefully to determine which, if any, are relevant to the specifications.

For example, MTE specification documents commonly include a footnote warning that the values are subject to change or modification without notice. Manufacturers do not generally modify existing MTE specifications unless significant changes in components or materials of construction warrant the establishment of new specifications. However, it may be necessary to contact the manufacturer to ensure that the appropriate MTE specification documents are obtained and applied.[31]

MTE specifications may state a recommended range of environmental operating conditions to ensure proper performance. They may also include a qualification indicating that all listed specifications are typical values referenced to standard conditions (e.g., 23 °C and 10 VDC excitation). This qualification implies that the primary performance specifications were developed from tests conducted under a particular set of conditions.

If so, additional specifications, such as thermal zero shift, thermal sensitivity shift and thermal transient response error, are included to account for the variation in actual MTE operating conditions from standard conditions. The MTE user must then consider whether or not these additional specifications are relevant to the MTE application.

9.7   Confidence Levels and Coverage Factors

Some MTE specifications are established by testing a selected sample of the manufacturer/model population. Since the test results are applied to the entire population, limits are developed to ensure that a large percentage of the produced items will perform as specified. Consequently, the specifications are confidence limits with associated confidence levels.

That is, the tolerance limits specified for an MTE performance parameter are established for a particular confidence level and degrees of freedom (or sample size). The confidence limits for values of a specific performance parameter, x, are expressed as

$$\pm L_x = \pm t_{\alpha/2,v} s_x \tag{9-3}$$

where

$$
\begin{aligned}
t_{\alpha/2,v} &= \text{t-statistic} \\
\alpha &= \text{significance level} = 1 - C/100 \\
C &= \text{confidence level (\%)} \\
v &= \text{degrees of freedom} = n - 1 \\
n &= \text{sample size}
\end{aligned}
$$

---

[31] i.e., the published specifications considered by the manufacturer to be applicable at the time the MTE was purchased.

$$s_x = \text{sample standard deviation.}$$

Ideally, confidence levels should be commensurate with what MTE manufacturers consider to be the maximum allowable false accept risk (FAR).[32] The general requirement is to minimize the probability of shipping an item or device with nonconforming (or out-of-tolerance) performance parameters. In this regard, the primary factor in setting the maximum allowable FAR may be the costs associated with shipping nonconforming products.

For example, an MTE manufacturer may require a maximum allowable FAR of 1% for all performance specifications. In this case, a 99% confidence level would be used to establish the MTE specification limits. Similarly, if the maximum allowable FAR is 5%, then a 95% confidence level should be used to establish the specification limits.

Unfortunately, manufacturers don't commonly report confidence levels for their MTE specifications. In fact, the criteria and motives used by manufacturers to establish MTE specifications are not often apparent. Most MTE manufacturers see the benefits, to themselves and their customers, of establishing specifications with high confidence levels. However, "specmanship" between MTE manufacturers can result in tighter specifications and increased out-of-tolerance occurrences.[33]

Alternatively, some manufacturers may test the entire MTE model population to ensure that all produced items are performing within specified limits prior to shipment. However, this compliance testing process does not ensure a 100% probability (or confidence level) that the customer will receive an in-tolerance item. The reason that 100% in-tolerance probability cannot be achieved is because of

1. Measurement uncertainty associated with the manufacturer MTE compliance testing process.

2. MTE bias drift or shift resulting from shock, vibration and other environmental extremes during shipping and handling.

Manufacturers may attempt to mitigate this problem by increasing the MTE specification limits. This can be accomplished by using a higher confidence level (e.g., 99.9%) to establish larger specification limits. Alternatively, some manufacturers may employ arbitrary guardbanding[34] methods and multiplying factors. In either case, the resulting MTE specifications are not equivalent to 100% confidence limits.

9.8   Probability Distributions

MTE performance parameters such as nonlinearity, repeatability, hysteresis, resolution, noise, thermal stability and zero shift constitute sources of measurement error. Measurement errors are random variables that can be characterized by probability distributions. Therefore, MTE performance parameters are also considered to be random variables that follow probability distributions.

---

[32] See NASA *Measurement Quality Assurance Handbook*, Annex 4 – *Estimation and Evaluation of Measurement Decision Risk.*

[33] See for example, Deaver, D.: "Having Confidence in Specifications."

[34] Guardbands are supplemental limits used to reduce false accept risk.

This concept is important to the interpretation and application of MTE specifications because an error distribution allows us to determine the probability that a performance parameter is in conformance or compliance with its specification. Error distributions include, but are not limited to normal, lognormal, uniform (rectangular), triangular, quadratic, cosine, exponential and u-shaped.

Some manufacturers may state that specification limits simply bound the value of an MTE parameter and do not imply any underlying probability distribution. Yet, when asked for clarification, many manufacturers indicate that the normal distribution is used as the underlying probability distribution.

For the sampled manufacturer/model described in Section 9.4, the performance parameter of an individual items may vary from the population mean. However, the majority of the items should perform well within their specification limits. As such, a central tendency exists that can be described by the normal distribution.

If the tolerance limits are asymmetric about a specified nominal value, it is still reasonable to assume that the performance parameter will tend to be distributed near the nominal value. In this case, the normal distribution may be applicable. However, the lognormal or other asymmetric probability distribution may be more applicable.

There are a couple of exceptions when the uniform distribution would be applicable. These include digital output resolution error and quantization error resulting from the digital conversion of an analog signal. In these instances, the specifications limits, $\pm L_{res}$ and $\pm L_{quan}$, would be 100% confidence limits defined as

$$\pm L_{res} = \pm \frac{h}{2} \tag{9-4}$$

and

$$\pm L_{quan} = \pm \frac{A}{2^{n+1}} \tag{9-5}$$

where

$$
\begin{aligned}
h &= \text{least significant display digit} \\
A &= \text{full scale range of analog to digital converter} \\
n &= \text{quantization significant bits.}
\end{aligned}
$$

9.9   Combining Specifications

In testing and calibration processes, an MTE performance parameter is identified as being in-tolerance or out-of-tolerance. In some cases, the tolerance limits are determined from a combination of MTE specifications. For example, consider the accuracy specifications for the DC voltage function of a digital multimeter listed in Table 9-4.

Table 9-4. DC Voltage Specifications for 8062A Digital Multimeter[35]

Specification	Value
200 mV Range Resolution	0.01 mV
200 mV Range Accuracy	0.05% of Reading + 2 digits

---

[35] Specifications from 8062A Instruction Manual downloaded from www.fluke.com

Specification	Value
2 V Range Resolution	0.1 mV
2 V Range Accuracy	0.05% of Reading + 2 digits
20 V Range Resolution	1 mV
20 V Range Accuracy	0.07% of Reading + 2 digits

For illustration purposes, it is assumed that a 5 V output is read with the DC voltage function of the multimeter. The appropriate accuracy specification would be $\pm$ (0.07% Reading + 2 mV).[36] To compute the combined accuracy specification, the % Reading must be converted to a value with units of mV.

$$0.07\% \text{ Reading} = (0.07/100) \times 5 \text{ V} \times 1000 \text{ mV/V} = 3.5 \text{ mV}$$

The total accuracy specification for the 5 V output reading would then be $\pm$ 5.5 mV. Combined accuracy specifications for different voltage ranges and output readings are summarized in Table 9-5 for comparison.

Table 9-5. DC Voltage Accuracy Specifications

Output Reading		200 mV Range	2 V Range	20 V Range
50 mV	0.05 V	$\pm$ 0.05 mV	$\pm$ 0.23 mV	$\pm$ 2.04 mV
100 mV	0.1 V	$\pm$ 0.07 mV	$\pm$ 0.25 mV	$\pm$ 2.07 mV
200 mV	0.2 V	$\pm$ 0.12 mV	$\pm$ 0.30 mV	$\pm$ 2.14 mV
500 mV	0.5 V		$\pm$ 0.45 mV	$\pm$ 2.35 mV
1000 mV	1 V		$\pm$ 0.70 mV	$\pm$ 2.70 mV
2000 mV	2 V		$\pm$ 1.20 mV	$\pm$ 3.40 mV
5000 mV	5 V			$\pm$ 5.50 mV
10000 mV	10 V			$\pm$ 9.00 mV

For another example, the combined tolerance limits for a Grade 2 gage block with 20 mm nominal length are computed. A subset of the data for different gage block grades published by NIST[37] is listed in Tables 9-6 and 9-7.

Table 9-6. Tolerance Grades for Metric Gage Blocks (μm)

Nominal	Grade .5	Grade 1	Grade 2	Grade 3
< 10 mm	0.03	0.05	+0.10, -0.05	+0.20, -0.10
< 25 mm	0.03	0.05	+0.10, -0.05	+0.30, -0.15
< 50 mm	0.05	0.10	+0.20, -0.10	+0.40, -0.20
< 75 mm	0.08	0.13	+0.25, -0.13	+0.45, -0.23
< 100 mm	0.10	0.15	+0.30, -0.15	+0.60, -0.30

---

[36] Understanding Specifications for Precision Multimeters, Application Note Pub_ID 11066-eng Rev 01, ©2006 Fluke Corporation.

[37] The Gage Block Handbook, NIST Monograph 180, 1995.

Table 9-7. Additional Tolerance for Length, Flatness, and Parallelism (μm)

Nominal	Grade .5	Grade 1	Grade 2	Grade 3
< 100 mm	± 0.03	± 0.05	± 0.08	± 0.10
< 200 mm		± 0.08	± 0.15	± 0.20
< 300 mm		± 0.10	± 0.20	± 0.25
< 500 mm		± 0.13	± 0.25	± 0.30

The appropriate specification limits for the 20 mm Grade 2 gage block are +0.10 μm, -0.05 μm and ± 0.08 μm. The first specification limits are asymmetric, while the second specification limits are symmetric. Consequently, the combined tolerance limits will be asymmetric and upper and lower tolerance limits (e.g., $+L_1$, $-L_2$) must be computed.

There are two possible ways to compute values for L_1 and L_2 from the specifications: linear combination or root-sum-square (RSS) combination.[38]

   1. Linear Combination

$$L_1 = 0.10 + 0.08 = 0.18$$
$$L_2 = 0.05 + 0.08 = 0.13$$

   2. RSS Combination

$$L_1 = \sqrt{(0.10)^2 + (0.08)^2} = \sqrt{0.0164} = 0.13$$
$$L_2 = \sqrt{(0.05)^2 + (0.08)^2} = \sqrt{0.0089} = 0.09$$

If the specifications are interpreted to be additive, then the combined tolerance limits for the 20 mm Grade 2 gage block are +0.18 μm, -0.13 μm. Alternatively, if they are to be combined in RSS, then the resulting tolerance limits are +0.13 μm, -0.09 μm.

---

[38] As illustrated in Section 9.2, linear and RSS combination cannot be used for MTE that have complex performance specifications.

CHAPTER 10: VALIDATING CONFORMANCE TO SPECIFICATIONS

Validating that MTE performance parameters are within specified tolerance limits is an important element of measurement quality assurance. In many instances, the validation of MTE performance is required by regulation, contract or policy directive.[39]

Companies and government organizations that maintain ISO 9001 or AS9100 quality management systems must also maintain an effective measuring and testing program by confirming that equipment are in conformance with manufacturer specifications.

Standards such as ISO 10012:2003 *Measurement Management Systems – Requirements for Measurement Processes and Measuring Equipment* contain quality assurance requirements for the metrological confirmation of MTE.[40] Confirmation of the performance parameters or attributes of commercial off-the-shelf (COTS) MTE is achieved through acceptance testing and periodic calibration.

10.1 Acceptance Testing

Acceptance tests are conducted to ensure that measuring equipment will function properly and meet performance requirements when operated in the application environment. Acceptance tests are conducted under simulated or actual operating conditions to determining the effectiveness and suitability of the MTE for its intended use.

Acceptance testing may not be required for COTS MTE when sufficient information about the accuracy, stability and overall reliability has already been established and the performance specifications are consistent with the intended operating conditions for the device. Of course, the COTS MTE parameters must be calibrated periodically to ensure that they are in conformance with the specified tolerance limits.

When MTE are specifically designed or modified for use in a new measurement application they should undergo acceptance testing to verify that

- The device is capable of measuring the desired quantities under the required operating conditions.

- The device meets established design requirements and is free of manufacturing defects.

- The device performs within its stated accuracy and other specification limits.

Acceptance tests should provide an explicit measure of the static and dynamic performance parameters and attributes of the device during exposure to applicable operating conditions and environments. These tests should be conducted by subject matter experts, in consultation with end users, to ensure that the equipment meets agreed-upon performance criteria.

---

[39] See for example, CxP 70059 *Constellation Program Integrated Safety, Reliability and Quality Assurance (SR&QA) Requirements* and NSTS 5300.4(1D-2) *Safety Reliability, Maintainability and Quality Provisions for the Space Shuttle Program.*

[40] Metrological confirmation consists of a set of operations that ensure that specified performance parameters and attributes comply with the requirements of the MTE's intended use.

10.1.1 Static Performance Characteristics
At a minimum, the static performance parameters listed below should be evaluated at baseline conditions by applying a range of steady-state inputs and recording the corresponding outputs.

- Zero Measured Output
- Deadband
- Full Scale Output
- Sensitivity
- Linearity
- Hysteresis
- Repeatability
- Zero Shift
- Sensitivity Shift

10.1.1.1 Zero Measured Output
The zero measured output or zero offset is established by measuring the output with no input signal applied. The warm-up time of the device is determined by measuring the zero output over a period of time (usually a few hours). If an external excitation source is applied, then it should also be monitored to ensure steady-state conditions are maintained during testing.

10.1.1.2 Deadband
The deadband of a device or instrument is established by slowly increasing the input signal in small increments until a non-zero output is observed. The deadband is the value of the input at which a non-zero output occurs.

10.1.1.3 Full Scale Output
The full scale or rated output is established by measuring the output when the full scale input is applied. Creep can be established by measuring changes in the full scale output during a specified period of time (usually several minutes to an hour). If an external excitation source is applied, then it should also be monitored to ensure steady-state conditions are maintained during testing.

10.1.1.4 Sensitivity
The sensitivity (or gain) of a device is evaluated by applying a range of steady-state inputs in an ascending or increasing direction, followed by a descending or decreasing direction. It is a good practice to apply at least two complete ascending and descending cycles, including at least five points in each direction (e.g., 0, 25, 50, 75, 100 percent of the full scale input). Sensitivity is characterized by the ratio of the output and input signals and is typically reported as a constant value over the input range.

10.1.1.5 Linearity
Linearity (or nonlinearity) is evaluated by applying a range of steady-state inputs in an ascending or increasing direction. Linearity is characterized by the maximum difference between the actual output response and the ideal linear response.[41]

---

[41] The ideal response may be determined from a specified sensitivity (i.e., slope), linear regression analysis or terminal (end point) method.

10.1.1.6 Hysteresis

Hysteresis is evaluated by applying a range of steady-state inputs in an ascending or increasing direction, followed by a descending or decreasing direction. It is a good practice to apply at least two complete ascending and descending cycles. Hysteresis is characterized by the maximum difference between ascending and descending outputs. In some cases, linearity and hysteresis are characterized as a single performance parameter.

10.1.1.7 Repeatability

Repeatability is characterized by applying a range of increasing (or decreasing) steady-state inputs for at least two cycles. These cycles must be applied under the same environmental conditions, using the same test equipment, procedures and personnel. The repeatability is the maximum difference in the output observed for a given input in the test range.

10.1.1.8 Zero and Sensitivity Shift

Shifts in zero offset and sensitivity are evaluated by observing any changes after applying two or more ascending and descending test cycles over a specified period of time. If an external excitation source is applied, then it should also be monitored to ensure steady-state conditions are maintained during testing.

10.1.2 Dynamic Performance Characteristics

At a minimum, the dynamic performance parameters listed below should be evaluated at baseline conditions by applying varying inputs and recording the corresponding outputs. The variation in input signal should reflect expected changes during actual device application (i.e., step change, ramp, sawtooth, sinusoidal).

- Frequency Response
- Phase Shift
- Resonance Frequency
- Damping Ratio
- Response Time
- Overshoot

The output response to changes in input should be recorded at a sufficient sampling rate to determine transient behavior. Transient stimulation sources or signal generators and high-speed data recorders are often employed for dynamic testing.

10.1.2.1 Frequency Response and Phase Shift

Frequency response is evaluated by applying a sinusoidal input signal over a range of frequencies and monitoring the corresponding sinusoidal output signal. The amplitude of the sinusoidal signal is set at three test points within the device range (e.g., 10%, 50%, 90%). The signal amplitude is held constant, while the frequency of the signal is increased from low to high frequency.

A reference device is used to monitor the input signal supplied to the unit under test (UUT).[42] The ratio of the amplitude of the UUT output to the amplitude of the input signal is computed at a series of frequencies. Frequency response is characterized by the change in the output/input

---

[42] The reference device must have a specified minimal amplitude loss or phase shift over the tested frequency range.

amplitude ratio, expressed as dB, over the frequency range. Amplitude flatness and ripple are characterized by how constant the output remains over the desired frequency range

At some point, as the frequency of the input signal is increased, the amplitude of the output signal will begin to decrease and the output signal will lag in time behind the input signal. Phase shift is characterized by the lag between the input sine wave and the output sine wave.

10.1.2.2 Resonance Frequency
The resonance of a device can be driven by mechanical, acoustic, electromagnetic, or other sources. Resonance frequency is characterized by the tendency of a device to oscillate at maximum amplitude at a certain frequency or frequencies. At these frequencies, even small periodic inputs can produce large variations in the output amplitude. When damping is small, the resonance frequency is approximately equal to the natural frequency of the device.

10.1.2.3 Damping Ratio
Damping ratio is evaluated during frequency response testing and is characterized by taking the frequency bandwidth at which the peak amplitude decreases by half (or 3 dB) and dividing by the natural or resonant frequency. The damping ratio increases with increased frequency bandwidth.

10.1.2.4 Response Time
Response time is evaluated by applying a rapid step change in the input. The step change can be an increase or decrease in the input. The response time is characterized by the time it takes for the output to reach a specified percentage of its final steady-state value.

10.1.2.5 Overshoot
Overshoot is evaluated during response time tests and is characterized as the maximum amount that the output signal exceeds its steady state output on its initial rise. Percent overshoot is computed from maximum value minus the step value divided by the step value.

10.1.3 Environmental Testing
Environmental tests are typically conducted under steady-state conditions using an atmospheric chamber or other climate control equipment. Atmospheric or environmental chambers are able to simulate extreme temperatures, humidity levels, altitude, radiation, and wind exposure. Shaker tables and centrifuge equipment are used to simulate vibration, shock and acceleration environments. In some cases, a track/rocket powered sled may be employed to simulate linear acceleration.

During environmental testing, the UUT is exposed to the controlled environmental conditions and allowed to come to steady state at a given input. Once equilibrium is reached, the output response is recorded. This procedure is repeated over a specified range of input values and environmental conditions. The following environmental effects are typically determined:

- Thermal Zero Shift
- Thermal Sensitivity Shift
- Humidity Effects
- Pressure or Altitude Error
- Acceleration Error
- Vibration and Shock Errors

The UUT should be exposed to each environmental condition and operated through at least one complete test cycle to establish the ability of the UUT to perform satisfactorily during each exposure. These tests should also verify the environmental operating ranges of the device. The testing duration for each environmental condition can range from minutes to hours, depending upon the intended UUT application.

Combinations of environmental conditions (e.g., temperature, vibration, acceleration) may more realistically represent the effects encountered during actual MTE operation. Therefore, combined environmental testing is recommended, when practical. Environmental testing standards and guidelines are discussed in Section 10.1.5.

10.1.3.1 Thermal Zero Shift
Thermal zero shift is determined by measuring the device output with no input signal applied for a range of environmental temperatures, while maintaining all other operating conditions at constant values. The thermal zero shift is characterized a the maximum, absolute change in the zero offset (from the baseline zero offset) divided by the temperature range tested.

10.1.3.2 Thermal Sensitivity Shift
Thermal sensitivity shift is determined by applying a range of steady-state inputs in an ascending or increasing direction for several different environmental temperatures, while maintaining all other operating conditions at constant values. The thermal sensitivity shift is characterized as the maximum, absolute change in the sensitivity (from the baseline sensitivity) divided by the temperature range tested.

10.1.3.3 Humidity Effects
The effects of humidity on zero offset and sensitivity are determined in a similar manner as described for the environmental temperature tests. The humidity effects are typically characterized as the maximum, absolute change (from baseline values) for the humidity range tested.

10.1.3.4 Pressure or Altitude Error
Pressure or altitude error is determined by applying a range of steady-state inputs in an ascending or increasing direction for several different environmental pressures. The pressure error is characterized as the maximum, absolute change in the zero offset and/or sensitivity (from the baseline values) for the pressure range tested.

10.1.3.5 Acceleration Error
Acceleration error is determined by applying a range of steady-state inputs in an ascending or increasing direction for a maximum specified acceleration or gravitational load. The acceleration error is characterized as the maximum, absolute change in the zero offset and/or sensitivity (from the baseline values) divided by the applied acceleration.

10.1.3.6 Vibration and Shock Error
Vibration error and shock error are determined by applying a range of inputs in an ascending or increasing direction for several different vibration amplitudes and frequencies. The vibration error is characterized as the maximum, absolute change in the zero offset and/or sensitivity (from the baseline values) for the vibration and frequency range tested. Shock error is characterized as the maximum, absolute change in zero offset and/or sensitivity (from baseline values) for the

maximum vibration and frequency tested.

10.1.4 Accelerated Life Testing

Acceptance testing primarily focuses on equipment performance during initial use and periodic calibration provides a means of tracking and maintaining equipment performance over time. However, when measuring devices are employed in aerospace or other long-term applications, periodic calibration may not be practical or possible. In such cases, the long-term reliability of these devices should be evaluated to determine if they can meet the desired performance requirements over their intended usage period.

Accelerated life testing (ALT) provides a means of evaluating how reliably a device can meet its performance specifications over an extended period of time. During ALT, the equipment or device is operated at stress conditions that will be encountered during actual use, but at a much quicker rate. In this regard, ALT methods are similar to the highly accelerated stress screening (HASS) method discussed in Chapter 4, but they are conducted using stress limits that are commensurate with actual use.

ALT should be designed to apply functional and environmental stresses to obtain the necessary performance data in an efficient, cost-effective manner. It is especially important to monitor and record the UUT performance parameters with sufficient frequency to ensure that pertinent changes are captured at relevant intervals throughout the duration of the test. The monitoring frequency will vary depending upon the type of test performed (e.g., static or dynamic).

The applied stresses can be constant, step-wise, ramped, cyclic, or random with respect to time. They should be commensurate with those typically encountered under actual usage conditions and based on the frequency or likelihood of occurrence of the stress event(s). Ideally, the applied stresses and stress limits used during the ALT should be based on measurements of the actual usage conditions. If these stresses or limits are unknown, preliminary testing of small sample sizes, using DOE methods, can be performed to ascertain the appropriate stresses and stress limits.

The ALT data are used to quantify the reliability of the device under the accelerated conditions. Mathematical modeling and analysis methods are then employed to provide an estimate of the life expectancy and long-term performance characteristics of the device under normal operating and environmental conditions. The test results can also be used to conduct risk assessments and equipment comparisons.

10.1.5 Testing Plans and Procedures

Standards and guidelines that provide acceptance and qualification testing methods and procedures for a variety of measuring equipment and devices can be obtained from professional organizations such as ISA, ASTM, ASME and IEC. Standards for the testing and reporting of MTE performance characteristics are listed in Appendix B.

There are no industry standards for environmental testing of MTE. However, many manufacturers test their products to MIL-STD-810F *Department of Defense Test Method Standard for Environmental Engineering Considerations and Laboratory Tests*. Additional environmental testing standards and handbooks are listed below.

- NASA SP-T-0023 Revision C, *Space Shuttle Specification Environmental Acceptance Testing*, 2001.

- MIL-HDBK-2036, *Department of Defense Handbook Preparation of Electronic Equipment Specifications*, 1999.

- IEC 60068-1, *Environmental Testing Part 1: General and Guidance*, 1988.

Also, there are no industry or government standards for accelerated life testing. Manufacturers develop their own test procedures based on documents such as MIL-HDBK-2164A, *Department of Defense Handbook Environmental Stress Screening Process for Electronic Equipment* and MIL-STD-810F. Additional, related documents are listed below.

- MIL-STD-202G, *Department of Defense Test Method Standard Electronic and Electrical Component Parts*, 2002.

- MIL-STD-781D *Reliability Testing for Engineering Development, Qualification and Production*, 1986.

10.2   Periodic Calibration

All equipment and devices used to obtain measurements must be properly calibrated to establish and maintain acceptable performance during use. ANSI/NCSLI Z540.3-2006 *Requirements for the Calibration of Measuring and Test Equipment* is the U.S. consensus standard that establishes the technical requirements for managing MTE and assuring that their calibrated parameters conform to specified performance requirements. This standard is applicable to organizations that use MTE to manufacture, modify, or test products, or to perform calibration services where measurement accuracy is important.

To achieve compliance with ANSI/NCSLI Z540.3, both quality control and assessment activities should be developed and implemented. These activities include

- Implementation of a documented process for the periodic calibration of MTE to verify that they are capable of accurately and reliably performing their intended measurement tasks.

- Assessment of measurement uncertainty for all equipment calibrations using appropriate analysis methods and procedures that account for all sources of uncertainty.[43]

MTE are periodically calibrated against measurement reference[44] to assess whether key performance parameters or attributes are within specified tolerance limits. More specifically, calibration is a testing process that measures MTE parameters or attributes and compares them to standard values to determine if they are in conformance or non-conformance with manufacturer or other specification limits.

MTE calibrations are often subsets of acceptance tests, in which MTE parameters are evaluated under less stressful operating conditions. The procedures vary depending upon the type of MTE being calibrated and the particular measurement scenario. The MTE being calibrated is typically

---

[43] Measurement uncertainty is discussed in Section 10.2.4.

[44] For example, mass or voltage standards, certified reference materials, or other MTE whose accuracy is traceable to the National Institute of Standards and Technology (NIST).

referred to as the UUT and the measurement reference is referred to as the reference standard.

Common calibration scenarios include:

1. The reference standard measures the value of the UUT attribute.
2. The UUT measures the value of the reference standard attribute.
3. The UUT and reference standard attribute values are measured with a comparator.
4. The UUT and reference standard both measure the value of an attribute of a common artifact.

In general, calibration procedures are developed to provide instructional guidance to ensure that calibration activities are performed in a manner consistent with the MTE application requirements. These procedural documents should include parameters or attributes to be calibrated, the operating range(s) tested, and the performance criteria used to establish compliance or non-compliance.

Calibration procedures often include steps for minor adjustments to bring MTE readings or outputs into better agreement with reference standard values. In this case, the MTE calibrating organization or facility should maintain records and reports containing as-found and as-left conditions. It is also a good practice to collect and store measured values and document the associated uncertainty estimates to ensure traceability to the national or international standard for that unit of measure.

ANSI/NCSL Z540.3 was developed to prescribe the requirements of an organization's calibration system including MTE inventory, calibration requirements, calibration intervals, and calibration procedures. ISO/IEC 17025:2005 *General Requirements for the Competence of Testing and Calibration Laboratories* is the consensus standard that contains the requirements that testing and calibration laboratories must meet to demonstrate that they operate a quality program or system and are capable of producing technically valid results. Guidance on the development, validation and maintenance of MTE calibration procedures can be found in NCSLI RP-3 *Calibration Procedures*.

The interval or period of time between calibrations may vary for each device depending upon the stability, application and degree of use. The establishment of calibration intervals is discussed in Section 10.4. ISO/IEC 17025:2005 and ANSI/NCSLI Z540.3 also require that the measurement results obtained during equipment calibration must be traceable to a national measurement institute, such as NIST. Measurement traceability is discussed in Section 10.2.5.

10.2.1 Laboratory Calibration

MTE are often calibrated in laboratories with controlled environmental conditions. The level of environmental control is largely dictated by the measurement reference(s) used to calibrate the UUT. The higher the reference standard level in the measurement traceability chain, the tighter the laboratory environmental control.

For example, expensive HVAC units are often employed to control temperature, humidity and dust in laboratories that use primary standards. Additional measures may also be undertaken to eliminate external vibration, microwave or radio frequency sources.

ISO/IEC 17025:2005, Section 5.3 *Accommodation and Environmental Conditions* states that "The laboratory shall ensure that the environmental conditions do not invalidate the results or adversely affect the required quality of any measurement." Section 5.3 further states that "The laboratory shall monitor, control, and record environmental conditions ... where they influence the quality of the results."

However, it is important that equipment used to monitor processing facilities or measure produced items be calibrated at actual usage conditions to ensure measurement quality is achieved. For example, a transducer used to monitor pipeline pressure in an oil refinery may be calibrated at several temperatures within its operating range to observe the effects on zero output and sensitivity.

As discussed in Section 10.1.3 environmental tests are typically conducted under steady-state conditions using an atmospheric chamber or other control equipment. In a calibration scenario, the reference standard(s) are typically isolated from the environmental conditions applied to the UUT.

Regardless of the external environmental conditions, the UUT should be calibrated at a sufficient number of points (including zero) in the appropriate operating range(s) to establish the conformance or non-conformance of the performance parameters to manufacturer specifications. It is a good practice to apply at least one complete ascending and descending cycle, including four or five points in each direction (e.g., 0, 25, 50, 75, 100 percent of the full scale input).

The laboratory calibration procedures should be of sufficient detail to enable qualified personnel[45] to properly calibrate the UUT. Laboratory calibration procedures typically include the following information:

- Description of the UUT, including manufacturer/model number, performance parameters calibrated, associated specification limits, and environmental conditions.

- Description of the reference standard(s) used, the associated specification limits and/or estimated uncertainties, and environmental conditions.

- Description of ancillary equipment used, the associated specification limits and/or estimated uncertainties, and environmental conditions (if applicable).

- Preliminary UUT inspection, operational and/or function[46] tests prior to calibration (if applicable).

- Detailed, step-by-step instructions for obtaining the calibration results.

- Instructions for recording, storing and analyzing the calibration data.

- Instructions for the labeling, storage and handling of calibrated device to maintain fitness for use.

Manufacturers of multifunction instruments often publish user guides or manuals that provide procedures for the calibration and functional testing of their equipment. However, most manufacturers do not have a policy of providing calibration procedures for their equipment.

---

[45] The qualifications of calibration and metrology personnel are discussed in Chapter 13.

[46] Function tests are quick checks designed to verify basic MTE operation.

Procedures for the laboratory calibration of MTE can be developed from various sources, including organizations such as ASTM, ISA, NIST and GIDEP.

ASTM and ISA procedures undergo peer review by technical experts on the respective standards committees. Procedures posted on the NIST and GIDEP websites do not undergo an approval process. In general, it is a good practice to consider all third party calibration procedures as basic guidance documents that require verification and validation prior to use.

Calibration laboratories often undergo a formal evaluation process to achieve accreditation to ISO/IEC 17025. This accreditation process is typically conducted annually to ensure that the laboratory meets the technical competence requirements of ISO/IEC 17025 as well as other requirements stipulated by laboratory accreditation bodies.[47]

10.2.2 In-Situ Calibration

In some cases, removing a measuring device to perform a laboratory calibration can be expensive and time-consuming. In process plant and pipeline applications, transducers are usually calibrated in a laboratory prior to installation, but they cannot readily be removed from their installation to be recalibrated. In aerospace applications, measurement systems installed for aircraft monitoring or flight testing must be calibrated in place to confirm that all components function properly independently and as a system.

In some instances, regulations may require the in-situ calibration of measuring equipment to ensure the device maintains its performance characteristics during operation. For example, custody transfer flowmeters are used to determine the amount of petroleum products transferred between companies, government agencies and other third-party entities. The millions of dollars associated with these measurement makes periodic in-situ flowmeter calibration a very important task.

In-situ calibration involves the testing or validation of a measuring device or system performed at its place of installation. This involves bringing the reference standard(s) to the installation location. Therefore, the reference standard(s) must be capable of being transported to the field location and maintaining its performance characteristics in this operating environment. Additional mechanical and/or electrical connections may also be required to allow the reference standard(s) to be installed in a manner that to ensures proper UUT calibration.

Unlike laboratory calibrations, in-situ calibrations are conducted in the actual operating environment and take installation effects into account. In-situ calibration should not to be confused with on-site calibration using a portable or mobile laboratory capable of performing tests or calibrations under controlled environmental conditions.

As with laboratory calibration, the environmental conditions must be monitored and recorded. The effects of the environment on the test or calibration results must also be evaluated and reported. The response of the reference standard(s) to environmental changes or other relevant operating conditions must also be known, documented and taken into account.

---

[47] In the U.S., accreditation bodies include the American Association of Laboratory Accreditation (A2LA) and the National Voluntary Association of Laboratory Programs (NAVLAP), among others.

In cases where it may not be practical or cost-effective to test a large number of points in the MTE operating range(s), a minimum set of tests are needed to establish the conformance or non-conformance of the key performance parameters to manufacturer specifications. If possible, it is a good practice to apply at least one complete ascending and descending cycle, including at least three points in each direction (e.g., 0, 50, 100 percent of the full scale input).

Some manufacturers sell portable calibration equipment to monitor selected performance parameters of the in-situ UUT. These portable devices have seen increased popularity because they are relatively easy and inexpensive to use. However, these portable calibrators are primarily designed to provide a quick check or verification of signal conditioning components. While some calibrators can check resistance, impedance or other internal electrical parameters, they are not capable of evaluating the actual physical or mechanical characteristics of most transducers.

An in-situ calibration is only achieved if one of the following occurs

1. The UUT and reference standard measure the same input and their outputs are compared and evaluated.

2. The reference standard applies a "known" input to the UUT and the UUT output is compared to manufacturer or other specified response information, such as prior calibration certificates or records.

The advantages of in-situ calibration include:

- Mechanical installation effects, electrical wiring and connectors, and signal conditioning are taken into account.

- Actual operating stresses are taken into account.

- Equipment downtime is minimized.

- Spare or replacement equipment requirements and associated costs are minimized.

The disadvantages of in-situ calibration include:

- Reference standards and ancillary equipment must be transferred to the field environment.

- Reference standards are exposed to field operating conditions.

- Environmental conditions during the calibration process do not reflect actual daily and seasonal variations that the UUT actually encounters.

- If transducer and signal conditioning components are calibrated together (e.g., system calibration), there is no way to determine if individual components are performing within manufacturer specifications.

The in-situ calibration procedures should be sufficiently detailed to enable qualified personnel to properly calibrate the UUT parameters. These procedures should contain the same information needed for laboratory calibrations, as listed in Section 10.2.1. It is also a recommended practice that the procedures include steps to perform in-situ calibrations for all measurement components, modules, sub-systems, and systems to ensure they conform to required performance specifications.

Some manufacturers may provide procedures for the in-situ calibration of their measuring equipment. However, standardized methods and procedures for in-situ calibration are not commonly published. In-situ calibration is not specifically addressed in ISO/IEC 17025, but some laboratory accrediting bodies stipulate requirements for laboratories that conduct in-situ or field calibrations.[48]

10.2.3 Self-Calibrating Equipment

Equipment with so-called "self-calibration," "self-diagnostic," or "self-adjusting" capabilities have become prevalent in recent years, especially in process measurement and control applications. Internal firmware routines are often included to detect and diagnose certain operating problems in a timely manner. Some equipment are designed to perform checks against built-in "standards" or references and make self-adjustments of the zero and span settings.

However, the inclusion of these features does not eliminate the need for periodic calibration using external, independent reference standards to ensure that the MTE parameters are in compliance with performance specifications. Despite what these features may imply, they only provide functional checks and basic diagnostics.

In general, the main problems with these devices include

- The built-in standards must be periodically calibrated against higher-level standards.

- The internal algorithms, calibration tables or curve fit equations must be routinely updated.

- The lack of measurement traceability.

10.2.4 Measurement Uncertainty

All measurements are accompanied by errors associated with the measurement equipment used, the environmental conditions during measurement, and the procedures used to obtain the measurement. Our lack of knowledge about the sign and magnitude of measurement error is called measurement uncertainty.

Therefore, the data obtained from MTE acceptance testing and calibration are not considered complete without statements of the associated measurement uncertainty. Using acceptance testing or calibration data, without accompanying uncertainty estimates, to establish conformance to manufacturer specifications will result in increased false accept or false reject risk.[49]

Since uncertainty estimates are used to support decisions, they should realistically reflect the measurement process. In this regard, the person tasked with conducting an uncertainty analysis must be knowledgeable about the overall measurement process, including the reference standard(s) and UUT. An in-depth coverage of key aspects of measurement uncertainty analysis and detailed procedures needed for developing such estimates is provided in NASA

---

[48] See for example, *C103 – General Checklist: Accreditation of Field Testing and Field Calibration Laboratories*, A2LA, May 2008.

[49] The application of uncertainty estimates for assessing the risk of falsely accepting or rejecting a calibrated parameter is discussed in Section 11.3.

10.2.5 Measurement Traceability

As previously discussed, MTE calibrations are typically performed with reference standards that are traceable to nationally or internationally recognized standards. National measurement or metrology laboratories, such as NIST in the United States, are responsible for developing, maintaining and disseminating the nationally recognized standards for basic, and many derived, measurement quantities. These national labs are also responsible for assessing the measurement uncertainties associated with the values assigned to these measurement standards.

For example, the NIST Policy on Traceability[50] requires the establishment of an unbroken chain of comparisons to stated reference standards.[51] Traceability is accomplished by ensuring that reference standards used to calibrate MTE are, in turn, calibrated by "higher level" reference standards and that this unbroken chain of calibrations is documented.

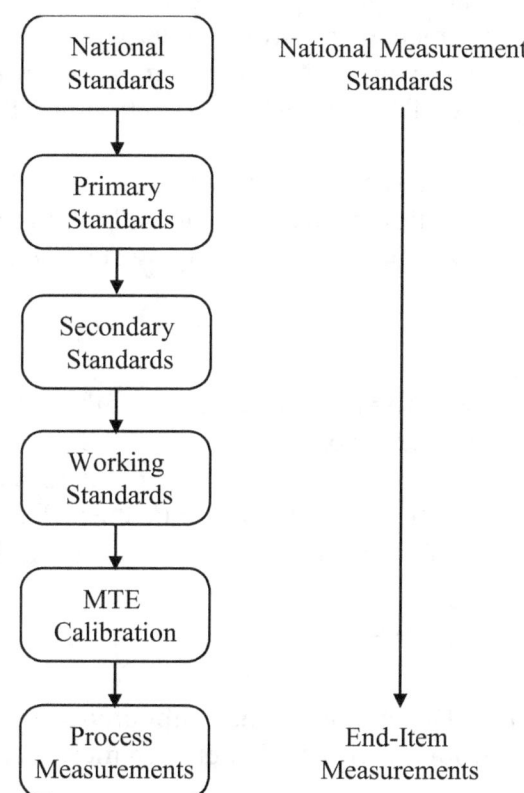

Figure 10-1 Measurement Traceability Chain

As shown in Figure 10-1, traceability is a hierarchical process that begins with standards from national laboratories (e.g., NIST, NPL, PTB) and ends with the MTE used to measure or monitor production processes. The traceability chain for the calibration of high accuracy MTE (e.g., primary standards) may only contain one link; whereas, the calibration of low accuracy MTE (e.g., end-item or process MTE) have many links and several reference standards. For example,

---

[50] See NIST Administrative Manual, Subchapter 5.16.

[51] A standard, generally having the highest metrological quality available at a given location or in a given organization, from which measurements there are derived (VIM).

thermocouple calibration must be traceable to the SI temperature scale and the unit of voltage.

At each link or level of the traceability chain, measurement errors are introduced that must be evaluated to yield an estimate of uncertainty. All measurement uncertainties propagated from the higher levels to the lower levels. Therefore, all relevant error sources must be identified and realistic uncertainty estimates must be made at each level. These uncertainties must be combined and unambiguously communicated to the next level in the calibration hierarchy.

Only measurement results and values of standards can be traceable. The internationally accepted definition states that traceability is "the property of the result of a measurement or the value of a standard whereby it can be related to stated references, usually national or international standards, through an unbroken chain of comparisons, all having stated uncertainties."

Consequently, measurement traceability is not truly accomplished until documentation for each link contains the assigned values of the calibrated MTE parameter or attribute, a stated uncertainty in the result, an uncertainty budget, the reference standards used in the calibration, and the specification of the environmental conditions under which the measurements were made. The allowable degradation in accuracy or increase in uncertainty is often specified for each link in the chain as a test accuracy ratio (TAR) or test uncertainty ratio (TUR).[52]

It is important to note that traceability is NOT the property of an instrument, calibration report or facility and cannot be achieved by following a particular calibration procedure or simply having an instrument calibrated by NIST without an accompanying statement of measurement uncertainty.

10.2.6 Conformance Testing

The primary purpose of calibration is to obtain an estimate of the value or bias of MTE attributes or parameters. The four calibration scenarios listed in Section 10.2 each yield an observed value referred to as a "measurement result" or "calibration result." In all scenarios, the calibration result is taken to be an estimation of the true UUT attribute bias. The relationship between the calibration result, δ, and the true UUT attribute bias, $e_{UUT,b}$, is generally expressed as

$$\delta = e_{UUT,b} + \varepsilon_{cal} \tag{10-1}$$

where ε_{cal} is the calibration error. Depending on the calibration scenario, δ can be a value that is directly measured or computed from the difference between measured values of the UUT and reference standard attributes.

Conformance testing assesses whether or not the value of δ falls within its specified tolerance limits.[53] If the value of δ falls outside of the specified tolerance limits for the UUT attribute, then it is typically deemed to be out-of-tolerance (OOT) or noncompliant. If a UUT attribute or parameter is found to be OOT, then an adjustment or maintenance action may be undertaken.

However, errors in the measurement process can result in an incorrect OOT assessment (false-

---

[52] The calculation and application of TAR and TUR are discussed in Chapter 11.

[53] Since the tolerance limits constitute the maximum permissible deviation or difference, it should be expressed in the units measured during calibration.

reject) or incorrect in-tolerance assessment (false-accept). Measurement process errors encountered in a given calibration scenario typically include:

$e_{std,b}$ = bias in the reference standard

e_{rep} = repeatability

e_{res} = resolution error

e_{op} = operator bias

e_{other} = other measurement error, such as that due to environmental corrections, ...
ancillary equipment variations, response to adjustments, etc.

The total measurement error is

$$\varepsilon_m = e_{std,b} + e_{rep} + e_{res} + e_{op} + e_{other} \tag{10-2}$$

For some calibration scenarios, ε_{cal} is synonymous with ε_m. In other scenarios, ε_{cal} and ε_m may not have equivalent sign or magnitude. In all calibration scenarios, the uncertainty in the estimation of $e_{UUT,b}$ (i.e., δ) is the uncertainty in the calibration error, u_{cal}.

For illustration, four calibration data points are shown in Figure 10-2 for a given MTE attribute. The vertical bars on each data point indicate $\pm u_{cal}$. Absent an assessment of u_{cal}, data points B, C and D would be considered in-tolerance and data point A out-of-tolerance. However, if uncertainty is included in the decision making process, then the compliance of data points C and D and the non-compliance of data point A are unclear. Only data point B can be considered in compliance with a high degree of confidence.

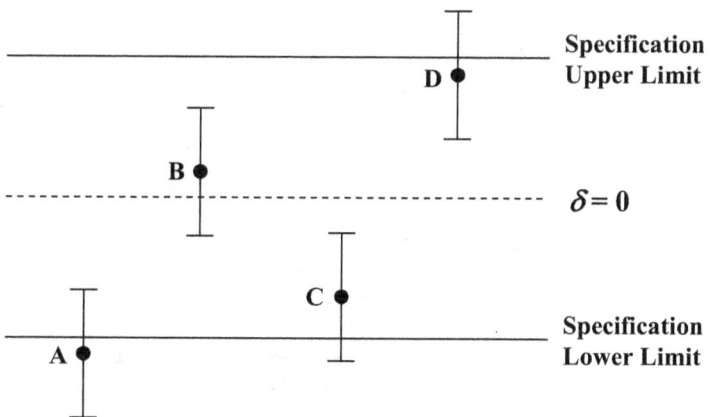

Figure 10-2 Four Calibration Results Relative to Specification Limits

To evaluate in-tolerance confidence levels, uncertainty estimates are required for each MTE calibration.[54] All relevant sources of measurement error must be identified and combined in a way that yields viable estimates. Recommended practices for using uncertainty estimates to evaluate in-tolerance probabilities for alternative calibration scenarios are given in NASA MQA Handbook – Annex 4 *Estimation and Evaluation of Measurement Decision Risk.*

---

[54] For example, see ASME B89.7.4.1-2005 *Measurement Uncertainty and Conformance Testing: Risk Analysis.*

10.2.7 Calibration Records

Records must be maintained on the calibration status, condition, and corrective actions or repairs for each MTE parameter. These calibration records must cover both conforming and non-conforming items and should have a suitable format, completeness, and detail to permit future analysis.

As-found and as-left in-tolerance or out-of-tolerance data[55] should be recorded for all MTE parameters tested. If as-found data indicate that all the MTE parameters are within specified tolerance limits, and no adjustments or corrections are made, then the as-left data should be the same as the as-found data.

It is also a recommended practice to record and analyze the actual measurement data to determine trends in the drift or shift of key MTE parameters. The measurement data obtained during calibration is often referred to as variables data. When variables data are required for analysis, the measurement results must have accompanying uncertainty estimates.

10.2.8 Calibration Intervals

MTE parameters are subject to errors arising from transportation, drift with time, use and abuse, environmental effects, and other sources. Consequently, the value or bias of an MTE parameter may increase, remain constant or decrease. The uncertainty in the MTE parameter value or bias, however, *always* increases with time since calibration.

Figure 10-3 illustrates uncertainty growth over time for a typical MTE parameter bias, ε_b. The sequence shows the probability distribution at three different times, with the uncertainty growth reflected in the spreads in the curves. The out-of-tolerance probabilities at the different times are represented by the shaded areas under the curves.

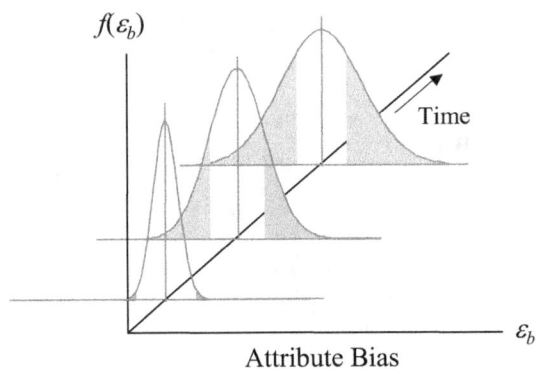

Figure 10-3 Measurement Uncertainty Growth.

The growth in uncertainty over time corresponds to an increased out-of-tolerance probability over time, or equivalently, to a decreased in-tolerance probability or measurement reliability $R(t)$ over time.

An effective MTE management program should set and achieve measurement reliability targets that ensure that MTE parameters are performing within specified tolerance limits over its calibration interval. These objectives are typically met by setting calibration intervals so that a

---

[55] i.e., the pre-calibration results and the post-calibration corrected or adjusted results.

high percentage of MTE parameters are observed to be in-tolerance when received for re-calibration.

Figure 10-4 illustrates the relationship between calibration intervals and end-of-period (EOP) measurement reliability targets, $R*$. The functional form of the reliability curve is dependent on the physical characteristics of the MTE, conditions and frequency of use, specification limits and calibration procedure.

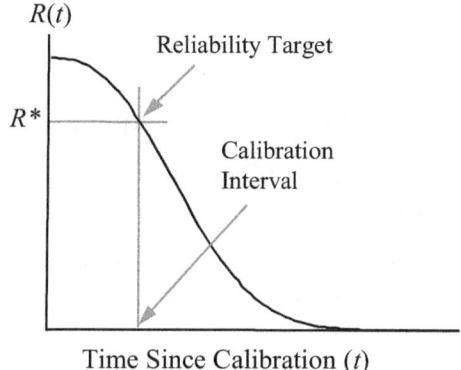

Figure 10-4 Measurement Reliability versus Time

In most cases, mathematical reliability models are established from historical in-tolerance or out-of-tolerance data obtained from the calibrations of a family or class of similar MTE.[56] Obviously, these reliability models are only valid if the population of MTE have similar conditions of use and specification limits, and are calibrated using comparable procedures.

Absent sufficient historical data, initial MTE calibration intervals are often based on the following criteria:

- Manufacturer recommended intervals
- Intervals for similar equipment
- Organizational policies
- Government directives or requirements
- Contract stipulations

The goal in establishing MTE calibration intervals is to meet measurement reliability requirements in a cost-effective manner. Recommended practices for establishing and maintaining optimal MTE calibration intervals are given in NASA MQA Handbook – Annex 5 *Establishment and Adjustment of Calibration Intervals.*

10.3 Bayesian Analysis
As previously discussed, calibration compares the value of an attribute or parameter of a calibrating device, such as a reference standard, against the UUT parameter value. A primary goal of measurement quality assurance is to maintain MTE accuracy over its usage life through periodic calibration or testing. However, periodic calibration of MTE used in aerospace applications or deployed in other remote locations can be difficult or impossible to achieve.

---

[56] For example, this would include groups containing the same MTE manufacturer/model or groups containing equivalent or substitute MTE.

88

The developers of measurement processes and equipment intended for long-duration space operations must employ design techniques and methods that provide the physical and functional characteristics to facilitate the evaluation of critical performance parameters throughout the MTE life or mission. Special in-situ calibration, self-diagnostics or interval extension schemes may be used, including

1. The use of carefully characterized astronomical artifacts as intrinsic measurement references.

2. The inclusion of internal reference standards with long-term accuracy commensurate with expected mission profiles.

3. The use of alternative or redundant measurement equipment for comparison.

4. The inclusion of firmware to conduct self-monitoring, function testing and diagnostics.

5. The establishment of tighter equipment tolerance limits to ensure conformance of performance parameters despite increasing uncertainty growth over time.

In traditional calibration, the reference standard is typically required to be inherently more accurate than the UUT. Accordingly, measurements made by the reference standard are given higher regard than measurements made by the UUT. If the calibration shows the UUT attribute to be out-of-tolerance, then the UUT is considered to be at fault.

In Bayesian analysis, measurement intercomparisons provide a means of estimating parameter biases and in-tolerance probabilities for both the UUT and reference standard alike, regardless of their relative accuracies. Statistical intercomparisons of measurements are made using two or more independent measuring attributes.[57] These measuring attributes may include both reference standards (and/or comparison devices) and UUTs.

Bayesian analysis is applicable under the following conditions:

- The measuring attributes are statistically independent.

- The measuring attributes exhibit enough variety to ensure that changes in attribute values are uncorrelated over the long term.

- The measuring attributes have been calibrated or tested before deployment.

- Drift or other uncertainty growth characteristics of the attributes have been defined before deployment.

The theory and application of Bayesian analysis for compliance or conformance testing are provided in NASA *Measurement Quality Assurance Handbook – Annex 4 Estimation and Evaluation of Measurement Decision Risk*.

10.4   Validation Reports
ISO/IEC 17025:2005 stipulates certain requirements regarding the reporting of calibration results: "The result of each test carried out by the lab shall be reported accurately, clearly, unambiguously, and objectively and in accordance with any specific instructions in the test

---

[57] A measuring attribute is regarded here as anything that has a nominal or stated value.

method. Where relevant, a statement of compliance/non-compliance with requirements and/or specifications should be included."

In general, MTE acceptance testing and calibration reports should include the following:

- Identification of the UUT manufacturer, model number and serial number.

- The purpose of the test or calibration.

- The UUT performance parameters and operating range(s) tested.

- A complete list of calibration equipment used, including manufacturer, model, serial number and date of last calibration for each device.

- The environmental and operating conditions applied to the UUT and the test sequence used.

- The environmental and operating conditions applied to the reference standards and other calibration equipment.

- The calibration procedures used or a document reference number.

- The measurement results and associated uncertainty estimates.

- Identification of out-of-tolerance parameters and the test conditions at which they occurred.

- An evaluation of the degree of correlation between laboratory test results and anticipated operating conditions during actual use.

- A summary of data analysis methods and procedures used or reference to external documentation.

CHAPTER 11: APPLYING SPECIFICATIONS

Manufacturer specifications are an important element of cost and quality control for testing, calibration and other measurement processes. They are used in the selection of MTE and the establishment of equivalent equipment substitutions for a given measurement application.

MTE specifications are used to estimate measurement uncertainty, establish tolerance limits for calibration and testing, and evaluate false accept risk and false reject risk. MTE parameters are periodically calibrated to determine if they are performing within manufacturer specified tolerance limits. In fact, the elapsed-time or interval between calibrations is often based on in-tolerance or out-of-tolerance data acquired from periodic calibrations.

11.1 Measurement Uncertainty Analysis

Manufacturer specifications can be used to conduct a preliminary assessment of the uncertainty in the nominal value or output of MTE. The analysis results can then be used to identify, and possibly mitigate, the largest contributors to overall uncertainty. These preliminary analyses should be conducted before MTE are selected or purchased.

For illustration, an uncertainty analysis is conducted for the load cell discussed in Chapter 9. This load cell has a nominal rated output of 2 mV/V for loads up to 5 lb$_f$. Therefore, the nominal sensitivity of the load cell is 0.4 mV/V/lb$_f$.

Based on the manufacturer specifications listed in Table 9-1, the load cell output will be affected by the following performance parameters:

- Excitation Voltage
- Nonlinearity
- Hysteresis
- Noise
- Zero Balance
- Temperature Effect on Output
- Temperature Effect on Zero

As discussed in Chapter 9, the appropriate transfer function or output equation for the load cell is expressed as

$$LC_{out} = \left[\left(W + TE_{out} \times TR_{°F} \right) \times S + NL + Hys + NS + ZO + TE_{zero} \times TR_{°F} \right] \times V_{Ex} \qquad (11\text{-}1)$$

where

$\quad LC_{out}$ = Output voltage, mV
$\quad W$ = Applied weight or load, lb$_f$
$\quad TE_{out}$ = Temperature effect on output, lb$_f$/°F
$\quad TR_{°F}$ = Temperature range, °F
$\quad S$ = Load cell sensitivity, mV/V/lb$_f$
$\quad NL$ = Nonlinearity, mV/V
$\quad Hys$ = Hysteresis, mV/V
$\quad NS$ = Noise and ripple, mV/V
$\quad ZO$ = Zero offset, mV/V
$\quad TE_{zero}$ = Temperature effect on zero, mV/V/°F

V_{Ex} = Excitation Voltage, V

Given some knowledge about the load cell parameters and their associated probability distributions, the uncertainty in the load cell output voltage can be estimated. A 3 lb_f applied load is used for this analysis.

Excitation Voltage (V_{Ex})
The load cell has an external power supply of 8 VDC excitation voltage with ± 0.25 V error limits. The excitation voltage error limits are interpreted to be a 95% confidence interval for a normally distributed error.

Nonlinearity (NL)
Nonlinearity is a measure of the deviation of the actual input-to-output performance of the load cell from an ideal linear relationship. Nonlinearity error is fixed at any given input, but varies with magnitude and sign over a range of inputs. Therefore, it is considered to be a random error that is normally distributed. The manufacturer specification of ± 0.05% of the rated output is interpreted to be a 95% confidence interval.

Hysteresis (Hys)
Hysteresis indicates that the output of the load cell is dependent upon the direction and magnitude by which the input is changed. At any input value, hysteresis can be expressed as the difference between the ascending and descending outputs. Hysteresis error is fixed at any given input, but varies with magnitude and sign over a range of inputs. Therefore, it is considered to be a random error that is normally distributed. The manufacturer specification of ± 0.05% of the rated output is interpreted to be a 95% confidence interval.

Noise (NS)
Noise is the nonrepeatability or random error intrinsic to the load cell that causes the output to vary from observation to observation for a constant input. This error source varies with magnitude and sign over a range of inputs and is normally distributed. The manufacturer specification of ± 0.05% of the rated output is interpreted to be a 95% confidence interval.

Zero Offset (ZO)
Zero offset occurs if the load cell generates a non-zero output for a zero applied load. Making an adjustment to reduce zero offset does not eliminate the associated error because there is no way to know the true value of the offset. The manufacturer specification of ± 1% of the rated output is interpreted to be a 95% confidence interval for a normally distributed error.

Temperature Effects (TE_{out} and TE_{zero})
Temperature can affect both the offset and sensitivity of the load cell. To establish these effects, the load cell is typically tested at several temperatures within its operating range and the effects on zero and sensitivity, or output, are observed.

The temperature effect on output of 0.005% load/°F specified by the manufacturer is equivalent to 0.00015 lb/°F for an applied load of 3 lb_f. The temperature effect on zero specification of 0.005% R.O./°F and the temperature effect on output are interpreted to be 95% confidence intervals for normally distributed errors.

92

The load cell is part of a tension testing machine, which heats up during use. The load cell temperature is monitored and recorded during the testing process and observed to increase from 75 °F to 85 °F. For this analysis, the 10 °F temperature range is assumed to have error limits of ± 2 °F with an associated 99% confidence level. The temperature measurement error is also assumed to be normally distributed.

The parameters used in the load cell output equation are listed in Table 11-1. The normal distribution is applied for all parameters.

Table 11-1. Parameters used in Load Cell Output Equation

Parameter Name	Description	Nominal or Stated Value	Error Limits	Confidence Level
W	Applied Load	3 lb$_f$		
S	Load Cell Sensitivity	0.4 mV/V/lb$_f$		
NL	Nonlinearity	0 mV/V	± 0.05% R.O. (mV/V)	95
Hys	Hysteresis	0 mV/V	± 0.05% R.O. (mV/V)	95
NS	Nonrepeatability	0 mV/V	± 0.05% R.O. (mV/V)	95
ZO	Zero Offset	0 mV/V	± 1% R.O. (mV/V)	95
$TR_{°F}$	Temperature Range	10 °F	± 2.0 °F	99
TE_{out}	Temp Effect on Output	0 lb$_f$/°F	± 0.005% Load/°F (lb$_f$/°F)	95
TE_{zero}	Temp Effect on Zero	0 mV/V /°F	± 0.005% R.O./°F (mV/V/°F)	95
V_{Ex}	Excitation Voltage	8 V	± 0.250 V	95

The error model for the load cell output is given in equation (11-2).

$$\varepsilon_{LC_{out}} = c_{NL}\varepsilon_{NL} + c_{Hys}\varepsilon_{Hys} + c_{NS}\varepsilon_{NS} + c_{ZO}\varepsilon_{ZO} + c_{TE_{out}}\varepsilon_{TE_{out}} + c_{TE_{zero}}\varepsilon_{TE_{zero}}$$
$$+ c_{TR_{°F}}\varepsilon_{TR_{°F}} + c_{V_{Ex}}\varepsilon_{V_{Ex}}$$

(11-2)

The coefficients (c_{NL}, c_{Hys}, etc.) are sensitivity coefficients that determine the relative contribution of the error sources to the total error in LC_{out}. The partial derivative equations used to compute the sensitivity coefficients are listed below. The measurement uncertainty analysis methods and procedures used in this analysis are provided in NASA *Measurement Quality Assurance Handbook* Annex 3 – *Measurement Uncertainty Analysis Principles and Methods*.

$$c_{NL} = \frac{\partial LC_{out}}{\partial NL} = V_{Ex} \qquad c_{Hys} = \frac{\partial LC_{out}}{\partial Hys} = V_{Ex}$$

$$c_{NS} = \frac{\partial LC_{out}}{\partial NS} = V_{Ex} \qquad c_{ZO} = \frac{\partial LC_{out}}{\partial ZO} = V_{Ex}$$

$$c_{TE_{out}} = \frac{\partial LC_{out}}{\partial TE_{out}} = TR_{°F} \times S \times V_{Ex} \qquad c_{TE_{zero}} = \frac{\partial LC_{out}}{\partial TE_{zero}} = TR_{°F} \times V_{Ex}$$

$$c_{TR\circ F} = \frac{\partial LC_{out}}{\partial TR\circ F} = (TE_{out} \times S + TE_{zero}) \times V_{Ex}$$

$$c_{V_{Ex}} = \frac{\partial LC_{out}}{\partial V_{Ex}} = (W + TE_{out} \times TR\circ F) \times S + NL + Hys + NS + ZO + TE_{zero} \times TR\circ F$$

Measurement uncertainty is the square root of the variance of the error distribution.[58] Therefore, the uncertainty in the load cell output is determined by applying the variance operator to equation (11-2).

$$u_{LC_{out}} = \sqrt{\mathrm{var}\left(\varepsilon_{LC_{out}}\right)}$$

$$= \sqrt{\mathrm{var}\left(\begin{array}{l} c_{NL}\varepsilon_{NL} + c_{Hys}\varepsilon_{Hys} + c_{NS}\varepsilon_{NS} + c_{ZO}\varepsilon_{ZO} + c_{TE_{out}}\varepsilon_{TE_{out}} \\ + c_{TE_{zero}}\varepsilon_{TE_{zero}} + c_{TR\circ F}\varepsilon_{TR\circ F} + c_{V_{Ex}}\varepsilon_{V_{Ex}} \end{array}\right)} \qquad (11\text{-}3)$$

There are no correlations between error sources in the load cell output equation. Therefore, the uncertainty in the load cell output can be expressed as

$$u_{LC_{out}} = \sqrt{\begin{array}{l} c_{NL}^2 u_{NL}^2 + c_{Hys}^2 u_{Hys}^2 + c_{NS}^2 u_{NS}^2 + c_{ZO}^2 u_{ZO}^2 + c_{TE_{out}}^2 u_{TE_{out}}^2 \\ + c_{TE_{zero}}^2 u_{TE_{zero}}^2 + c_{TR\circ F}^2 u_{TR\circ F}^2 + c_{V_{Ex}}^2 u_{V_{Ex}}^2 \end{array}} \qquad (11\text{-}4)$$

The uncertainty in the load cell output is computed from the uncertainty estimates and sensitivity coefficients for each load cell parameter.

As previously discussed, all of the error sources identified in the load cell output equation are assumed to follow a normal distribution. Therefore, the corresponding uncertainties are estimated from the error limits, $\pm L$, confidence level, p, and the inverse normal distribution function, $\Phi^{-1}(\cdot)$.

$$u = \frac{L}{\Phi^{-1}\left(\dfrac{1+p}{2}\right)} \qquad (11\text{-}5)$$

For example, the uncertainty in the excitation voltage error is estimated to be

$$u_{V_{Ex}} = \frac{0.25\ \mathrm{V}}{\Phi^{-1}\left(\dfrac{1+0.95}{2}\right)} = \frac{0.25\ \mathrm{V}}{1.9600} = 0.1276\ \mathrm{V}.$$

The sensitivity coefficients are computed using the parameter nominal or mean values.

---

[58] The basis for the mathematical relationship between error and uncertainty is presented in Chapter 2 of NASA *Measurement Quality Assurance Handbook* Annex 3 – *Measurement Uncertainty Analysis Principles and Methods*.

$$c_S = \left(W + TE_{out} \times TR_{\circ F}\right) \times V_{Ex}$$
$$= \left(3\,\text{lb}_f + 0\,\text{lb}_f / {\circ}\text{F} \times 10\,{\circ}\text{F}\right) \times 8\,\text{V}$$
$$= 3\,\text{lb}_f \times 8\,\text{V}$$
$$= 24\,\text{lb}_f \times \text{V}$$

$$c_{NL} = V_{Ex} \qquad c_{Hys} = V_{Ex} \qquad c_{NS} = V_{Ex} \qquad c_{ZO} = V_{Ex}$$
$$= 8\,\text{V} \qquad\quad = 8\,\text{V} \qquad\quad = 8\,\text{V} \qquad\quad = 8\,\text{V}$$

$$c_{TR_{\circ F}} = \left(TE_{out} \times S + TE_{zero}\right) \times V_{Ex} \qquad c_{TE_{out}} = TR_{\circ F} \times S \times V_{Ex}$$
$$= \left(0 \times 0.4\,\text{mV/V/lb}_f + 0\right) \times 8\,\text{V} \qquad\quad = 10\,{\circ}\text{F} \times 0.4\,\text{mV/V/lb}_f \times 8\,\text{V}$$
$$= 0 \qquad\qquad\qquad\qquad\qquad\qquad\quad = 32\,{\circ}\text{F} \times \text{mV/lb}_f$$

$$c_{TE_{zero}} = TR_{\circ F} \times V_{Ex}$$
$$= 10\,{\circ}\text{F} \times 8\,\text{V}$$
$$= 80\,{\circ}\text{F} \times \text{V}$$

$$c_{V_{Ex}} = \left(W + TE_{out} \times TR_{\circ F}\right) \times S + NL + Hys + NS + ZO + TE_{zero} \times TR_{\circ F}$$
$$= \left(3\,\text{lb}_f + 0\,\text{lb}_f/{\circ}\text{F} \times 10\,{\circ}\text{F}\right) \times 0.4\,\text{mV/V/lb}_f + 0\,\text{mV/V} + 0\,\text{mV/V} + 0\,\text{mV/V} + 0\,\text{mV/V}$$
$$+ 0\,\text{mV/V/}{\circ}\text{F} \times 10\,{\circ}\text{F}$$
$$= 3\,\text{lb}_f \times 0.4\,\text{mV/V/lb}_f$$
$$= 1.2\,\text{mV/V}$$

The estimated uncertainties and sensitivity coefficients for each parameter are listed in Table 11-2.

Table 11-2. Estimated Uncertainties for Load Cell Parameters

Param. Name	Nominal or Stated Value	± Error Limits	Conf. Level	Standard Uncertainty	Sensitivity Coefficient	Component Uncertainty
W	$3\,\text{lb}_f$					
S	$0.4\,\text{mV/V/lb}_f$				$24\,\text{lb}_f \times \text{V}$	
NL	$0\,\text{mV/V}$	$\pm 0.001\,\text{mV/V}$	95	$0.0005\,\text{mV/V}$	$8\,\text{V}$	$0.0041\,\text{mV}$
Hys	$0\,\text{mV/V}$	$\pm 0.001\,\text{mV/V}$	95	$0.0005\,\text{mV/V}$	$8\,\text{V}$	$0.0041\,\text{mV}$
NS	$0\,\text{mV/V}$	$\pm 0.001\,\text{mV/V}$	95	$0.0005\,\text{mV/V}$	$8\,\text{V}$	$0.0041\,\text{mV}$
ZO	$0\,\text{mV/V}$	$\pm 0.02\,\text{mV/V}$	95	$0.0102\,\text{mV/V}$	$8\,\text{V}$	$0.0816\,\text{mV}$
$TR_{\circ F}$	$10\,{\circ}\text{F}$	$\pm 2.0\,{\circ}\text{F}$	99	$0.7764\,{\circ}\text{F}$	0	
TE_{out}	$0\,\text{lb}_f/{\circ}\text{F}$	$\pm 1.5 \times 10^{-4}\,\text{lb}_f/{\circ}\text{F}$	95	$7.65 \times 10^{-5}\,\text{lb}_f/{\circ}\text{F}$	$32\,{\circ}\text{F} \times \text{mV/lb}_f$	$0.0024\,\text{mV}$
TE_{zero}	$0\,\text{mV/}{\circ}\text{F}$	$\pm 0.0001\,\text{mV/V/}{\circ}\text{F}$	95	$5 \times 10^{-5}\,\text{mV/V/}{\circ}\text{F}$	$80\,{\circ}\text{F} \times \text{V}$	$0.0041\,\text{mV}$
V_{Ex}	$8\,\text{V}$	$\pm 0.25\,\text{V}$	95	$0.1276\,\text{V}$	$1.2\,\text{mV/V}$	$0.1531\,\text{mV}$

The component uncertainties listed in Table 11-2 are the products of the standard uncertainty and sensitivity coefficient for each parameter. The nominal load cell output is computed to be

$$LC_{out} = W \times S \times V_{Ex} = 3 \text{ lb}_f \times 0.4 \text{ mV/V/lb}_f \times 8 \text{ V} = 9.6 \text{ mV}.$$

The total uncertainty in the load cell output is computed by taking the root sum square of the component uncertainties.

$$u_{LC_{out}} = \sqrt{\begin{array}{l} (0.0041 \text{ mV})^2 + (0.0041 \text{ mV})^2 + (0.0041 \text{ mV})^2 + (0.0816 \text{ mV})^2 \\ + (0.0024 \text{ mV})^2 + (0.0041 \text{ mV})^2 + (0.1531 \text{ mV})^2 \end{array}}$$

$$= \sqrt{0.0302 \text{ mV}^2} = 0.174 \text{ mV}$$

The total uncertainty is equal to 1.8% of the 9.60 mV load cell output. The Welch-Satterthwaite formula[59] is used to compute the degrees of freedom for the uncertainty in the load cell output voltage, as shown in equation (11-6).

$$v_{LC_{Out}} = \frac{u_{LC_{out}}^4}{\displaystyle\sum_{i=1}^{8} \frac{c_i^4 u_i^4}{v_i}} \tag{11-6}$$

where

$$\sum_{i=1}^{8} \frac{c_i^4 u_i^4}{v_i} = \frac{c_{NL}^4 u_{NL}^4}{v_{NL}} + \frac{c_{Hys}^4 u_{Hys}^4}{v_{Hys}} + \frac{c_{NS}^4 u_{NS}^4}{v_{NS}} + \frac{c_{ZO}^4 u_{ZO}^4}{v_{ZO}} + \frac{c_{TR\circ F}^4 u_{TR\circ F}^4}{v_{TR\circ F}} + \frac{c_{TE_{out}}^4 u_{TE_{out}}^4}{v_{TE_{out}}}$$
$$+ \frac{c_{TE_{zero}}^4 u_{TE_{zero}}^4}{v_{TE_{zero}}} + \frac{c_{V_{Ex}}^4 u_{V_{Ex}}^4}{v_{V_{Ex}}}$$

The degrees of freedom for all of the error source uncertainties are assumed infinite. Therefore, the degrees of freedom for the uncertainty in the load cell output is also infinite.

The Pareto chart, shown in Figure 11-1, indicates that the excitation voltage and zero offset errors are, by far, the largest contributors to the overall uncertainty in the load cell output. Replacement of the power supply with a precision voltage source could significantly reduce the total uncertainty in the load cell output. Mitigation of the zero offset error, however, would probably require a different model load cell.

The results of this analysis show that manufacturer specifications can be used to estimate the expected uncertainty in the load cell output and to identify the major contributors to this uncertainty.

---

[59] A discussion and derivation of the Welch-Satterthwaite formula are given in NASA *Measurement Quality Assurance Handbook* Annex 3 – *Measurement Uncertainty Analysis Principles and Methods*.

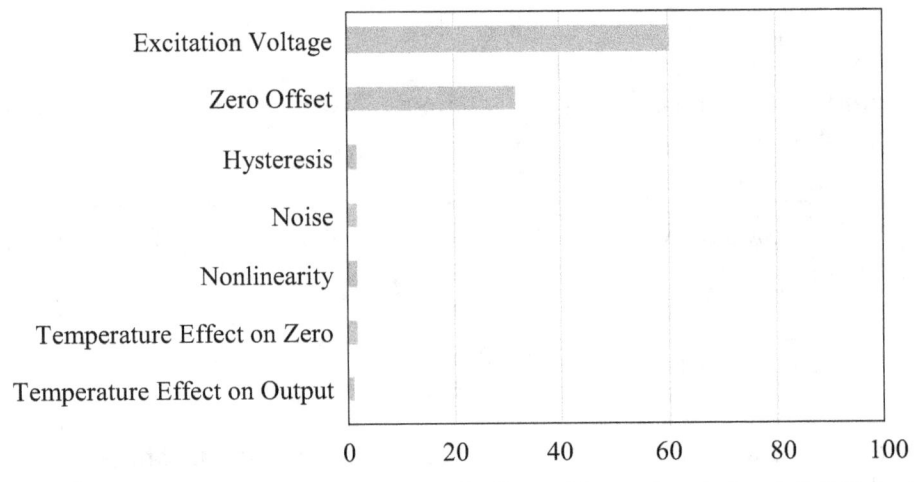

Percent Contribution to Uncertainty in Load Cell Output

Figure 11-1 Pareto Chart for Uncertainty in Load Cell Output

11.2 Tolerance Limits

The load cell output uncertainty and degrees of freedom can be used to compute confidence limits that are expected to contain the output voltage with some specified confidence level or probability, p. The confidence limits are expressed as

$$LC_{out} \pm t_{\alpha/2,v} \times u_{LC_{out}} \qquad (11\text{-}7)$$

where the multiplier, $t_{\alpha/2v}$, is a t-statistic and $\alpha = 1 - p$ is the significance level. Tolerance limits for a 95% confidence level (i.e., $p = 0.95$) are computed using a corresponding t-statistic of $t_{0.025,\infty} = 1.96$.

$$9.60\,\text{mV} \pm 1.96 \times 0.174\,\text{mV} \ \text{or} \ 9.60\,\text{mV} \pm 0.341\,\text{mV}$$

The uncertainty analysis and tolerance limit calculation processes can be repeated for different applied loads. The results are summarized in Table 11-3.

Table 11-3. Uncertainty in Load Cell Output Voltage versus Applied Load

Applied Load	Output Voltage	Total Uncertainty	Confidence Limits (95%)
1 lb$_f$	3.2 mV	0.097 mV	$\pm$ 0.190 mV
2 lb$_f$	6.4 mV	0.131 mV	$\pm$ 0.257 mV
3 lb$_f$	9.6 mV	0.174 mV	$\pm$ 0.341 mV
4 lb$_f$	12.8 mV	0.220 mV	$\pm$ 0.431 mV
5 lb$_f$	16.0 mV	0.268 mV	$\pm$ 0.525 mV

The *Slope* and *Intercept* functions of the Microsoft Excel application can then be used to obtain a linear fit of the above tolerance limits as a function of the applied load. These calculations yield an intercept value of 0.095 mV and a slope value of 0.084 mV/lb$_f$. From these values, the 95% confidence limits can be written as

$$\pm (0.095\,\text{mV} + 0.084\,\text{mV/lb}_f \times \text{Applied Load}).$$

As shown in Figure 11-1, the uncertainty in the excitation voltage has a significant impact on the confidence limits computed in this load cell example. These limits are also influenced, to a lesser extent, by the temperature range that the load cell is exposed to during use.

11.3   Measurement Decision Risk Analysis

The probability of making an incorrect decision based on a measurement result is called measurement decision risk. ANSI/NCSLI Z540.3:2006, Section 5.2 states that "b) Where calibrations provide for verification that measurement quantities are within specified tolerances, the probability that incorrect acceptance decisions (false accept) will result from calibration tests shall not exceed 2% and shall be documented."

Various probability concepts and definitions are employed in the computation of measurement decision risk. For instance, the probability that an MTE parameter or attribute accepted during calibration and testing as being in-tolerance is actually out-of-tolerance (OOT) is called false accept risk (FAR). Conversely, the probability that an MTE parameter determined to be OOT is actually in-tolerance is called false reject risk (FRR).

Measurement decision risk concepts for estimating in-tolerance probabilities of MTE parameters and alternative decision criteria are briefly discussed in this section. An in-depth coverage of the methods and principles of measurement decision risk analysis and the estimation and evaluation of FAR and FRR are provided in NASA *Measurement Quality Assurance Handbook,* Annex 4 – *Estimation and Evaluation of Measurement Decision Risk.*

As discussed in Section 10.2.6, the primary purpose of calibration is to obtain an estimate of the value or bias of MTE attributes or parameters. Another important purpose is to ascertain the conformance or non-conformance of MTE parameters with specified tolerance limits. The calibration result, δ, is taken to be an estimation of the true parameter bias, $e_{UUT,b}$, of the unit under test (UUT). The relationship between δ and $e_{UUT,b}$ is generally expressed as

$$\delta = e_{UUT,b} + \varepsilon_{cal} \qquad (11\text{-}8)$$

where ε_{cal} is the calibration error.

If the value of δ falls outside of the specified tolerance limits for the UUT parameter, then it is typically deemed to be OOT. However, errors in the calibration process can result in an incorrect OOT assessment (false-reject) or incorrect in-tolerance assessment (false-accept).

The probability that the UUT parameter is in-tolerance is based on the calibration result and its associated uncertainty. All relevant calibration error sources must be identified and combined in a way that yields viable uncertainty estimates.

For illustration, the calibration results for an individual (i.e., serial numbered) load cell selected from the manufacturer/model population described in Section 11.2 will be evaluated. The confidence limits listed in Table 11-3 for this manufacturer/model will be used to assess the probability that the measured voltage output from the load cell is in-tolerance.

In this example, the load cell is calibrated using a weight standard that has a stated value of 3.02 lb$_f$ and expanded uncertainty of ± 0.01 lb$_f$. During calibration, an excitation voltage of 8 VDC ± 0.25 V is supplied to the load cell. The calibration weight is connected and removed from the load cell several times to obtain repeatability data. The load cell output voltage is measured using a digital multimeter. The manufacturer's published accuracy and resolution specifications for the DC voltage function of the digital multimeter[60] are listed in Table 11-4.

Table 11-4. DC Voltage Specifications for 8062A Multimeter

Specification	Value	Units
200 mV Range Resolution	0.01	mV
200 mV Range Accuracy	0.05% of Reading + 2 digits	mV

The applicable calibration error sources are listed below.

- Bias in the value of the calibration weight, ε_W
- Excitation voltage error, $\varepsilon_{V_{Ex}}$
- DC voltmeter digital resolution, $\varepsilon_{V_{res}}$
- DC voltmeter accuracy, $\varepsilon_{V_{acc}}$
- Repeat measurements error, $\varepsilon_{V_{rep}}$

Weight Standard (W)
The 3.02 lb$_f$ weight standard has an expanded uncertainty of ± 0.01 lb$_f$. In this analysis, these limits are interpreted to represent a coverage factor, k, equal to 2. The associated error distribution is characterized by the normal distribution.

Excitation Voltage (V_{Ex})
The ± 0.25 V excitation voltage error limits are interpreted to be a 95% confidence interval for a normally distributed error.

DC Voltmeter Resolution (V_{res})
The digital display resolution is specified as 0.01 mV. Therefore, the resolution error limits are ± 0.005 mV and are interpreted to be the minimum 100% containment limits for this uniformly distributed error source.

DC Voltmeter Accuracy (V_{acc})
The overall accuracy of the DC Voltage reading for a 0 to 200 mV range is specified as ± (0.05% of reading + 2 digits). These error limits are interpreted to be a 95% confidence interval for normally distributed errors.

Repeatability (V_{rep})
The error resulting from repeat measurements can result from various physical phenomena such as temperature variation or the act of removing and re-suspending the calibration weight multiple times. Uncertainty due to repeatability error will be estimated from the standard deviation of the measurement data listed in Table 11-5.

---

[60] Specifications from 8062A Instruction Manual downloaded from www.fluke.com

Table 11-5. DC Voltage Readings

Repeat Measurement	Measured Output Voltage (mV)	Offset from Nominal Output[61] (mV)
1	9.85	0.19
2	9.80	0.14
3	9.82	0.16
4	9.84	0.18
5	9.80	0.14
Average	9.82	0.16
Std. Dev.	0.023	0.023

The load cell calibration equation is given in equation (11-9).

$$LC_{cal} = W \times S \times V_{Ex} + V_{res} + V_{acc} + V_{rep} \qquad (11\text{-}9)$$

Nominal values and error limits for the parameters used in the load cell calibration output equation are listed in Table 11-6. The normal distribution is applied for all parameters except V_{res}, which has a uniform distribution.

Table 11-6. Parameters used in Load Cell Calibration Output Equation

Param. Name	Description	Nominal or Average Value	Error Limits	Confidence Level
W	Applied Load	3.02 lb$_f$	$\pm\,0.01$ lb$_f$	95.45
S	Load Cell Sensitivity	0.4 mV/V/lb$_f$		
V_{Ex}	Excitation Voltage	8 V	$\pm\,0.250$ V	95
V_{res}	Voltmeter Resolution	0 mV	$\pm\,0.005$ mV	100
V_{acc}	Voltmeter Accuracy	0 mV	$\pm\,(0.05\%\ \text{Rdg} + 0.02\text{mV})$	95
V_{rep}	Repeatability	0.16 mV		

The error model for the load cell calibration is given in equation (11-10).

$$\varepsilon_{cal} = c_W \varepsilon_W + c_{V_{Ex}} \varepsilon_{V_{Ex}} + c_{V_{res}} \varepsilon_{V_{res}} + c_{V_{acc}} \varepsilon_{V_{acc}} + c_{V_{rep}} \varepsilon_{V_{rep}} \qquad (11\text{-}10)$$

The uncertainty in the load cell calibration is determined by applying the variance operator to equation (11-10).

$$u_{cal} = \sqrt{\text{var}\left(\varepsilon_{cal}\right)}$$
$$= \sqrt{\text{var}\left(c_W \varepsilon_W + c_{V_{Ex}} \varepsilon_{V_{Ex}} + c_{V_{res}} \varepsilon_{V_{res}} + c_{V_{acc}} \varepsilon_{V_{acc}} + c_{V_{rep}} \varepsilon_{V_{rep}}\right)} \qquad (11\text{-}11)$$

There are no correlations between error sources, so the uncertainty in the load cell calibration can

---

[61] In the analysis of the calibration results, the nominal load cell output = 3.02 lb$_f \times$ 0.4 mV/V/lb$_f \times$ 8 V = 9.66 mV.

be expressed as

$$u_{cal} = \sqrt{c_W^2 u_W^2 + c_{V_{Ex}}^2 u_{V_{Ex}}^2 + c_{V_{res}}^2 u_{V_{res}}^2 + c_{V_{acc}}^2 u_{V_{acc}}^2 + c_{V_{rep}}^2 u_{V_{rep}}^2} \qquad (11\text{-}12)$$

The uncertainty in the load cell output is computed from the uncertainty estimates and sensitivity coefficients for each parameter. The partial derivative equations used to compute the sensitivity coefficients are listed below.

$$c_W = \frac{\partial LC_{out}}{\partial W} = S \times V_{Ex} \qquad\qquad c_{V_{Ex}} = \frac{\partial LC_{out}}{\partial V_{Ex}} = W \times S$$
$$= 0.4 \text{ mV/V/lb}_f \times 8 \text{ V} \qquad\qquad = 3.02 \text{ lb}_f \times 0.4 \text{ mV/V/lb}_f$$
$$= 3.2 \text{ mV/lb}_f \qquad\qquad\qquad = 1.21 \text{ mV/V}$$

$$c_{V_{res}} = \frac{\partial LC_{out}}{\partial V_{res}} = 1 \qquad c_{V_{acc}} = \frac{\partial LC_{out}}{\partial V_{acc}} = 1 \qquad c_{V_{rep}} = \frac{\partial LC_{out}}{\partial V_{rep}} = 1$$

The uncertainty in the bias of the weight standard is estimated to be

$$u_W = \frac{0.01 \text{ lb}_f}{\Phi^{-1}\left(\dfrac{1+0.9545}{2}\right)} = \frac{0.01 \text{ lb}_f}{2} = 0.005 \text{ lb}_f.$$

The uncertainty due to excitation voltage error is estimated to be

$$u_{V_{Ex}} = \frac{0.25 \text{ V}}{\Phi^{-1}\left(\dfrac{1+0.95}{2}\right)} = \frac{0.25 \text{ V}}{1.9600} = 0.1276 \text{ V}.$$

The uncertainty due to voltmeter digital resolution is estimated to be

$$u_{V_{res}} = \frac{0.005 \text{ mV}}{\sqrt{3}} = \frac{0.005 \text{ mV}}{1.732} = 0.0029 \text{ mV}.$$

The uncertainty due to voltmeter accuracy is estimated to be

$$u_{V_{acc}} = \frac{\left(9.658 \text{ mV} \times \dfrac{0.05}{100} + 0.02 \text{ mV}\right)}{\Phi^{-1}\left(\dfrac{1+0.95}{2}\right)}$$
$$= \left(\frac{0.0048 \text{ mV} + 0.02 \text{ mV}}{1.9600}\right) = \frac{0.0248 \text{ mV}}{1.9600} = 0.0127 \text{ mV}.$$

The uncertainty due to repeatability in the load cell voltage measurements is equal to the standard deviation of the sample data.

$$u_{V_{rep}} = 0.023 \, \text{mV}$$

The sample mean is the quantity of interest in this analysis. Therefore, the uncertainty in the mean value due to repeatability should be included in the calculation of total uncertainty in the load cell output voltage. The uncertainty in the mean value is defined as

$$u_{\bar{V}_{rep}} = \frac{u_{V_{rep}}}{\sqrt{n}} \qquad (11\text{-}13)$$

where n is the sample size. The uncertainty in the mean value is estimated to be

$$u_{\bar{V}_{rep}} = \frac{0.023 \, \text{mV}}{\sqrt{5}} = \frac{0.023 \, \text{mV}}{2.236} = 0.0103 \, \text{mV}.$$

The estimated uncertainties and sensitivity coefficients for each parameter are summarized in Table 11-7.

Table 11-7. Estimated Uncertainties for Load Cell Calibration Output

Param. Name	Nominal or Stated Value	± Error Limits	Conf. Level	Standard Uncertainty	Sensitivity Coefficient	Component Uncertainty
W	3.02 lb$_f$	± 0.01 lb$_f$	99	0.005 lb$_f$	3.2 mV/lb$_f$	0.0160 mV
V_{Ex}	8 V	± 0.25 V	95	0.1276 V	1.21 mV/V	0.1544 mV
V_{acc}	0 mV	± 0.0248 mV	95	0.0127 mV	1	0.0127 mV
V_{res}	0 mV	± 0.005 mV	100	0.0029 mV	1	0.0029 mV
$u_{\bar{V}_{rep}}$	0.16 mV			0.0103 mV	1	0.0103 mV

The total uncertainty in the load cell calibration output is computed by taking the root sum square of the component uncertainties.

$$u_{cal} = \sqrt{(0.0160 \, \text{mV})^2 + (0.1544 \, \text{mV})^2 + (0.0127 \, \text{mV})^2 + (0.0029 \, \text{mV})^2 + (0.0103 \, \text{mV})^2}$$

$$= \sqrt{0.0244 \, \text{mV}^2} = 0.156 \, \text{mV}$$

The Pareto chart, shown in Figure 11-2, indicates that excitation voltage is the largest contributor to the overall uncertainty. The uncertainties due to the weight standard, voltmeter accuracy, repeatability and voltmeter resolution provide much lower contributions to the overall uncertainty.

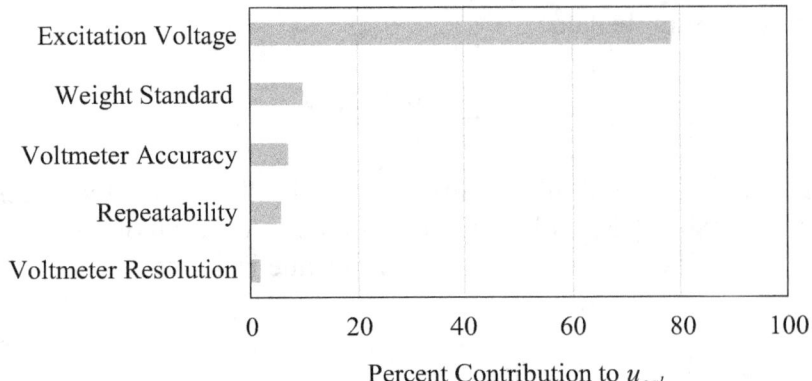

Figure 11-2 Pareto Chart for Load Cell Calibration Uncertainty

The Welch-Satterthwaite formula is used to compute the degrees of freedom for the uncertainty in the load cell output voltage.

$$v_{cal} = \frac{u_{cal}^4}{\dfrac{c_W^4 u_W^4}{v_W} + \dfrac{c_{V_{Ex}}^4 u_{V_{Ex}}^4}{v_{V_{Ex}}} + \dfrac{c_{V_{acc}}^4 u_{V_{acc}}^4}{v_{V_{acc}}} + \dfrac{c_{V_{res}}^4 u_{V_{res}}^4}{v_{V_{res}}} + \dfrac{c_{\overline{V}_{rep}}^4 u_{\overline{V}_{rep}}^4}{v_{\overline{V}_{rep}}}} \tag{11-14}$$

Since the uncertainty in the excitation voltage is the major contributor to the total uncertainty, the degrees of freedom for the total uncertainty is infinite.

$$v_{cal} = \frac{(0.156)^4}{\dfrac{(0.0160)^4}{\infty} + \dfrac{(0.1544)^4}{\infty} + \dfrac{(0.0127)^4}{\infty} + \dfrac{(0.0029)^4}{\infty} + \dfrac{(0.0103)^4}{4}}$$

$$= \infty$$

The difference between the average measured load cell voltage output and the nominal or expected output for an applied load of 3.02 lb$_f$ is

$$\overline{\delta} = (9.82 - 9.66)\,\text{mV} = 0.16\,\text{mV}$$

where $\overline{\delta}$ is an estimate of the bias in the load cell output, $e_{UUT,b}$, at the time of calibration. The confidence limits for the load cell bias can be expressed as

$$\overline{\delta} \pm t_{\alpha/2,v} \times u_{cal} \tag{11-15}$$

For a 95% confidence level, $t_{0.025,\infty} = 1.9600$ and the confidence limits for $e_{UUT,b}$ are computed to be

$$0.16\,\text{mV} \pm 1.96 \times 0.156\,\text{mV} \ \text{ or } \ 0.16\,\text{mV} \pm 0.31\,\text{mV}.$$

11.3.1 In-tolerance Probability

Figure 11-3 shows the $\varepsilon_{UUT,b}$ probability distribution for the population of manufacturer/model load cells. The spread of the distribution is based on the specification tolerance limits computed in Section 11.2 for a 3 lb$_f$ applied load. The calibration result, $\bar{\delta}$, provides an estimate of the unknown value of $\varepsilon_{UUT,b}$ for the individual load cell.

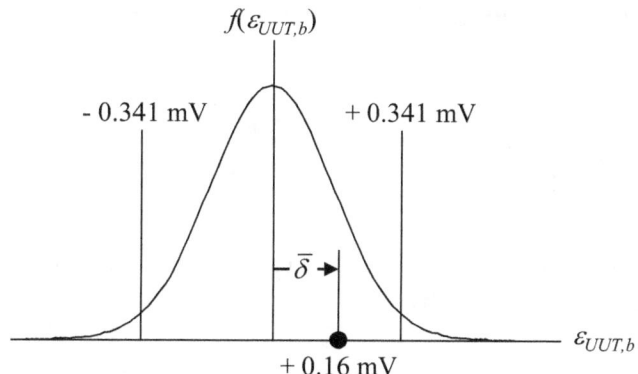

Figure 11-3 Load Cell Bias Distribution – 3 lb$_f$ Input Load

Given the value of $\bar{\delta} = 0.16\,\text{mV}$ observed during calibration, it appears that the load cell output is in-tolerance. However, in deciding whether the load cell output is in-tolerance or not, it is important to consider that $\bar{\delta}$ is also affected by the bias in the calibration weight, ε_W, the bias in the excitation voltage, $\varepsilon_{V_{Ex}}$, and the bias in the digital voltmeter reading, $\varepsilon_{V_{acc}}$. Consequently, the actual bias in the load cell output voltage may be larger or smaller than 0.16 mV.

While the value of $\varepsilon_{UUT,b}$ for the calibrated load cell is unknown, there is a 95% confidence that it is contained within the limits of $0.16\,\text{mV} \pm 0.31\,\text{mV}$. Figure 11-4 shows the probability distribution for $\varepsilon_{UUT,b}$ centered around $\bar{\delta} = 0.16\,\text{mV}$. The black bar depicts the ± 0.31 mV confidence limits and the shaded area depicts the probability that $\varepsilon_{UUT,b}$ falls outside of the ± 0.341 mV manufacturer specification limits.

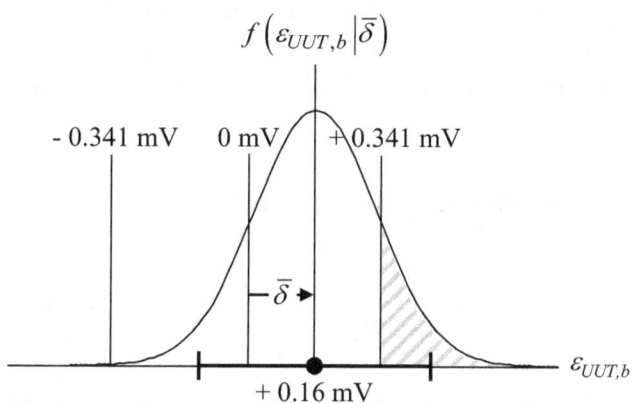

Figure 11-4 OOT Probability of Calibrated Load Cell

While the OOT probability is lower than the in-tolerance probability, it may introduce a significant risk of falsely accepting a non-conforming parameter. Bayesian analysis methods are employed to estimate UUT parameter biases and compute in-tolerance probabilities based on *a priori* knowledge and on measurement results obtained during testing or calibration.

Prior to calibration, the uncertainty in the UUT parameter bias, $u_{UUT,b}$, is estimated from the probability distribution for $\varepsilon_{UUT,b}$, the specification limits and the associated confidence level (*a priori* in-tolerance probability). First, the in-tolerance probability prior to calibration is computed by integrating the distribution function, $f(\varepsilon_{UUT,b})$, between the specified $\pm L$ tolerance limits

$$P(in) = \int_{-L}^{L} f(e_{UUT,b}) de_{UUT,b} = 2\Phi\left(\frac{L}{u_{UUT,b}}\right) - 1 \tag{11-16}$$

where Φ is the normal distribution function. The uncertainty in the UUT parameter bias $u_{UUT,b}$ is solved for by rearranging equation (11-16).

$$u_{UUT,b} = \frac{L}{\Phi^{-1}\left(\dfrac{1 + P(in)}{2}\right)} \tag{11-17}$$

where Φ^{-1} is the inverse normal distribution function.

After calibration, the values of δ and u_{cal} are used to estimate the UUT parameter bias.

$$\beta = \frac{u_{UUT,b}^2}{u_A^2} \delta \tag{11-18}$$

where

$$u_A = \sqrt{u_{UUT,b}^2 + u_{cal}^2} \tag{11-19}$$

From equations (11-18) and (11-19), it can be seen that the Bayesian estimate of the UUT parameter bias will be less than or equal to calibration result, δ. For example, if the values of $u_{UUT,b}$ and u_{cal} are equal, then $\beta = \delta / 2$. Conversely if u_{cal} is much smaller than $u_{UUT,b}$, then $\beta \cong \delta$.

The uncertainty in the Bayesian estimate of the UUT parameter bias is

$$u_\beta = \frac{u_{UUT,b}}{u_A} u_{cal} \tag{11-20}$$

Finally, the post-calibration in-tolerance probability for the UUT parameter is computed from equation (11-21).

$$P(in) = \Phi\left(\frac{L + \beta}{u_\beta}\right) + \Phi\left(\frac{L - \beta}{u_\beta}\right) - 1 \tag{11-21}$$

The 95% confidence limits for a given manufacturer/model load cell were computed and summarized in Table 11-3. These limits were computed using uncertainty analysis methods to combine the manufacturer specifications for different applied loads. From equation (11-17), the uncertainty in the load cell bias prior to calibration is estimated to be

$$u_{UUT,b} = \frac{0.341 \text{ mV}}{\Phi^{-1}\left(\dfrac{1+0.95}{2}\right)} = \frac{0.341 \text{ mV}}{1.9600} = 0.174 \text{ mV}.$$

This bias uncertainty is equivalent to the standard deviation of the probability distribution for the population of manufacturer/model load cells, shown in Figure 11-3.

The calibration results of an individual load cell, for a 3 lb$_f$ applied load, were also analyzed. Using the calibration results of $\bar{\delta} = 0.16 \text{ mV}$ and $u_{cal} = 0.156 \text{ mV}$, the bias in the load cell output is computed from equations (11-18) and (11-19).

$$\begin{aligned}
\beta &= \frac{(0.174 \text{ mV})^2}{(0.174 \text{ mV})^2 + (0.156 \text{ mV})^2} \times 0.16 \text{ mV} \\
&= \frac{0.0303 \text{ mV}^2}{0.0546 \text{ mV}^2} \times 0.16 \text{ mV} \\
&= 0.554 \times 0.16 \text{ mV} \\
&= 0.09 \text{ mV}
\end{aligned}$$

The uncertainty in this bias is computed from equation (11-20).

$$\begin{aligned}
u_\beta &= \frac{0.174 \text{ mV}}{0.234 \text{ mV}} \times 0.156 \text{ mV} \\
&= 0.746 \times 0.156 \text{ mV} \\
&= 0.116 \text{ mV}
\end{aligned}$$

Finally, equation (11-21) is used to compute the probability that the load cell output is in-tolerance during calibration.

$$\begin{aligned}
P(in) &= \Phi\left(\frac{0.341 \text{ mV} + 0.09 \text{ mV}}{0.116 \text{ mV}}\right) + \Phi\left(\frac{0.341 \text{ mV} - 0.09 \text{ mV}}{0.116 \text{ mV}}\right) - 1 \\
&= \Phi\left(\frac{0.404}{0.116}\right) + \Phi\left(\frac{0.251}{0.116}\right) - 1 \\
&= \Phi(3.483) + \Phi(2.164) - 1 \\
&= 0.99975 + 0.9848 - 1 \\
&= 0.985 \text{ or } 98.5\%.
\end{aligned}$$

The risk of falsely accepting the load cell output as in-tolerance is

$$FAR = 1 - P(in)$$
$$= 1 - 0.985$$
$$= 0.015 \text{ or } 1.5\%.$$

Applying Bayesian methods to analyze pre- and post-calibration information provides an explicit means of estimating the in-tolerance probability of MTE parameters. In most cases, the single variable with the greatest impact on measurement decision risk is the *a priori* in-tolerance probability of the UUT parameter.[62] Unfortunately, many manufacturers don't report in-tolerance probabilities or confidence levels for their MTE specifications. In such cases, a simplified relative accuracy criterion is often used to control measurement decision risk.

11.3.2 Relative Accuracy Criterion

Historically, the control of measurement decision risk has been embodied in requirements specifying the relative accuracy of the test or calibration process to the accuracy of the UUT parameter or attribute being tested or calibrated.[63,64] The common practice has been to require that the relative ratio of the accuracy of the UUT parameter to the collective accuracy of the calibration standards (expressed as uncertainty) must be at least 4 to 1 (4:1).

The 4:1 test accuracy ratio (TAR) requirement means the specified tolerance for the UUT parameter must be greater than or equal to four times the combined uncertainties of all the reference standards used in the calibration process. The effectiveness of this risk control requirement is debatable, in part, because of the lack of agreement regarding the calculation of TAR.

A more explicit relative accuracy requirement has been defined in ANSI/NCSL Z540.3:2006. This standard states that where it is not practicable to estimate FAR, the test uncertainty ratio (TUR) shall be equal to or greater than 4:1. TUR is defined as the ratio of the span of the UUT tolerance limits to twice the 95% expanded uncertainty of the measurement process used for calibration. TUR differs from TAR in the inclusion of all pertinent measurement process errors.

$$\text{TUR} = \frac{L_1 + L_2}{2U_{95}}$$

(11-22)

where U_{95} is equal to the expanded uncertainty of the measurement process and is computed by multiplying u_{cal} by a coverage factor, k_{95}, that is expected to correspond to a 95% confidence level.

$$U_{95} = k_{95}u_{cal}$$

(11-23)

---

[62] See Section 5.1 of NASA *Measurement Quality Assurance Handbook* Annex 4 – *Estimation and Evaluation of Measurement Decision Risk.*

[63] MIL-STD 45662A, *Calibration Systems Requirements,* U.S. Dept. of Defense, 1 August 1988.

[64] ANSI/NCSL Z540.1 (R2002), *Calibration Laboratories and Measuring and Test Equipment – General Requirements,* July 1994.

In ANSI/NCSL Z540.3:2006, $k_{95} = 2$ and TUR is valid only in cases where the two-sided tolerance limits are symmetric (i.e., $L_1 = L_2$). Therefore, the UUT tolerance limits would be expressed in the form $\pm L$, and TUR expressed as

$$\text{TUR} = \frac{L}{U_{95}} \qquad (11\text{-}24)$$

While the TUR $\geq$ 4:1 requirement is intended to provide some loose control of false accept risk, it doesn't establish whether the UUT parameter is in- or out-of-tolerance with the specification limits. For example, the TUR for the load cell calibration example is computed to be

$$\begin{aligned}
\text{TUR} &= \frac{0.341\,\text{mV}}{2 \times 0.156\,\text{mV}} \\
&= \frac{0.341\,\text{mV}}{0.312\,\text{mV}} \\
&= 1.09.
\end{aligned}$$

While this TUR fails to meet the $\geq$ 4:1 requirement, the corresponding FAR calculated in Section 11.2.1 meets the $\leq$ 2% requirement. In this case, TUR may provide some figure of merit about the calibration process, but it doesn't provide an assessment of FAR.

Since the *a priori* in-tolerance probability of the UUT parameter is crucial for the evaluation of measurement decision risk, the TUR 4:1 criterion is not considered to be a true "risk control" method. So, when should the 4:1 TUR criterion be used? The 4:1 TUR should only be used when absolutely no information about the *a priori* in-tolerance probability is available.

11.4   Calibration Interval Analysis

Periodic calibration comprises a significant cost driver in the life cycle of MTE. It also provides a major safeguard in controlling uncertainty growth and reducing the risk of substandard MTE performance during use. The goal in establishing MTE calibration intervals is to meet measurement reliability requirements in a cost-effective manner.

As discussed in Section 10.2.8, MTE parameters are subject to errors arising from transportation, drift with time, use and abuse, environmental effects, and other sources. Consequently, the MTE parameter bias may increase, remain constant or decrease, but the uncertainty in this bias always increases with time since calibration.

The growth in uncertainty over time corresponds to an increased out-of-tolerance probability over time, or equivalently, to a decreased in-tolerance probability or measurement reliability over time. In most cases, mathematical reliability models are established from historical in-tolerance or out-of-tolerance data obtained from the calibrations of a family of manufacturer/model MTE or a group of similar MTE. Absent sufficient historical data, initial MTE calibration intervals are often based on manufacturer recommended intervals.

11.4.1 Reliability Modeling

The primary objective of reliability modeling is the establishment of calibration intervals that ensure that appropriate measurement reliability targets are met. A reliability model predicts the in-tolerance probability for the parameter bias population as a function of time elapsed since measurement. It can be thought of as a function that quantifies the *stability* of the manufacturer/model population.

The measurement reliability of the parameter bias at time t is

$$R(t) = \int_{-L_1}^{L_2} f[\varepsilon_b(t)]d\varepsilon_b \qquad (11\text{-}25)$$

where $f[\varepsilon_b(t)]$ is the probability density function (pdf) for the parameter bias $\varepsilon_b(t)$ and $-L_1$ and L_2 are the specification tolerance limits. For example, if $\varepsilon_b(t)$ is assumed to be normally distributed, then

$$f[\varepsilon_b(t)] = \frac{1}{\sqrt{2\pi}u(t)}e^{-[\varepsilon_b(t)-\mu(t)]/2u^2(t)} \qquad (11\text{-}26)$$

where $u(t)$ is the bias uncertainty and $\mu(t)$ represents the expected or true parameter bias at time t. The relationship between L_1, L_2, $\varepsilon_b(t)$ and $\mu(t)$ is shown in Figure 11-5, along with the bias distribution for the MTE parameter of interest. The measurement reliability $R(t)$ is equal to the in-tolerance probability.

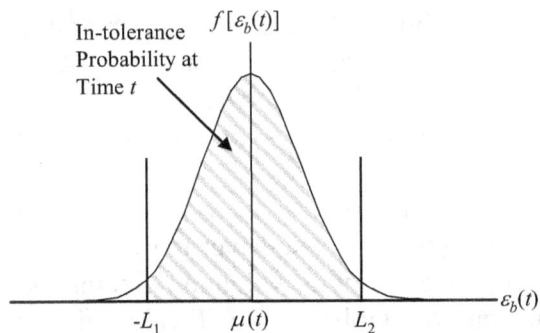

Figure 11-5 Parameter Bias Distribution.

A reliability model is defined by a mathematical equation that describes how measurement reliability changes over time. A calibration interval analysis application can be used to determine the reliability model that "best fits" the calibration history data and to compute the corresponding model coefficients.

If a reliability modeling application is not available, then an applicable reliability model must be chosen based on knowledge about the stability of the MTE parameter over time. Descriptions of commonly used reliability models and guidance on the selection and application of these reliability models can be found in the NASA *Measurement Quality Assurance Handbook* Annex 5 – *Establishment and Adjustment of Calibration Intervals*.

Once the reliability model has been established it can be used to identify the appropriate calibration interval for a desired reliability target, as shown in Figure 11-6. A measurement reliability target is determined by the requirements for calibration accuracy and is usually referenced to the end of the calibration interval or to a value averaged over the duration of the calibration interval.

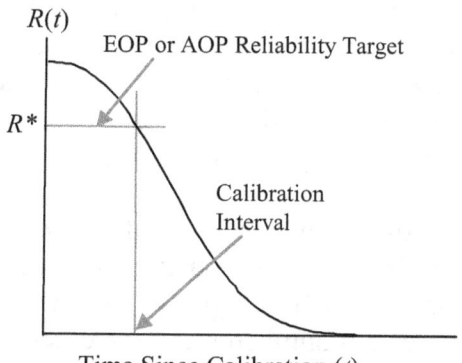

Figure 11-6 Measurement Reliability versus Time.

The establishment of end-of-period (EOP) or average-over-period (AOP) reliability targets involves a consideration of several trade-offs between the desirability of controlling measurement uncertainty growth and the cost associated with maintaining such control.

In practice, many organizations have found it expedient to manage measurement reliability at the instrument rather than the parameter level. In these cases, an item of MTE is considered out-of-tolerance if one or more of its parameters is found to be out-of-tolerance.

11.4.2 Manufacturer Specified Intervals

Calibration intervals specified by MTE manufacturers are often developed from the analysis of stability data at the parameter level. The following information is needed to implement the specified calibration interval:

- The parameter tolerance limits.

- The period of time over which the parameter values will be contained within the tolerance limits.

- The probability that the MTE parameter will be contained within the tolerance limits for the specified period of time.

Unfortunately, manufacturers may not communicate all of the necessary information to adequately adopt their interval recommendations. In this case, supporting calibration data from similar equipment, engineering design analysis and manufacturer expertise can be helpful in establishing initial calibration intervals.

CHAPTER 12: FIRMWARE AND SOFTWARE-BASED SPECIFICATIONS

MTE with automated self-testing, self-adjustment and self-calibration capabilities have become prevalent in recent years, especially in process measurement and control applications. Internal firmware routines are often included to detect and diagnose certain operating problems in a timely manner. Some equipment are designed to perform checks of their settings against built-in "standards" or references and make self-adjustments of the zero and span settings.

The primary rationale for this approach is to monitor and control the MTE performance characteristics and parameters via firmware rather than by manual adjustments, trimming or component selection. MTE controlled by firmware can be configured or upgraded to add new capability to existing hardware. Some manufacturers have extended this concept by providing remote software downloads via the Internet that allow customers to add options and features to their existing equipment.

12.1 Background

Firmware is the embedded software that integrates internal components, functions and user interface of the measuring device. Firmware consists of machine language instructions or algorithms that are executed by a microprocessor or microcontroller to monitor, set and/or adjust the configuration and functional settings of the device. The self-calibration and self-adjustment algorithms used in the firmware are typically developed during MTE design and modified during final article testing.

Firmware is commonly stored on read-only memory (ROM) or erasable programmable read-only memory (EPROM) embedded in the device. Many of these devices can be updated by replacing the ROM or EPROM chips. In some cases, the instructions on the EPROM can be updated using manufacturer supplied software.

The shift to embedded computing capabilities in instruments and other MTE has been driven by the demand for increased functionality and flexibility as well as the low cost and size of microprocessors. The addition of firmware provides a means for synchronizing, integrating and controlling the various features of the measuring device. Most complex instruments also provide a minimum amount of self-testing and automated calibration.

12.2 Advantages

Performance characteristics and parameters for each serial numbered item can be specified in the firmware. These factory installed (programmed) specifications can also be customized to suit a particular measurement application through custom factory calibration. This approach moves away from the generic manufacturer specifications to distinct performance parameters for a given hardware and firmware configuration.

Additional advantages include:

- User configuration and customization for a particular measurement application

- Automation of complex or repetitive measurement tasks

- Calculations to convert raw measurement data to desired output units

- Control of measurement components such as ADC or DAC, filters, and output displays

- Built-in self-testing, diagnostics and failure condition alerts

- Built-in measurement algorithms and numerical calculations

12.3   Disadvantages

The primary disadvantage arises from the built-in calibration and correction features that are designed to perform checks of configuration settings, compare performance parameters against built-in "standards" or references and make self-adjustments of the zero and span settings. Despite what these "auto-cal" features or functions may imply, they only provide functional checks and basic diagnostics. More importantly, to achieve measurement traceability, the built-in standards must be periodically calibrated against higher-level standards.

12.4   Limitations and Problems

Major problems can arise from the lack of transparency regarding the role and function of MTE firmware and software. Built-in standards or references may not be readily accessible for calibration or testing. In addition, MTE users may not be able to update or modify built-in calibration factors or algorithms.

In many cases, the MTE performance parameters are established by an algorithm located in firmware or software. However, if manufacturers don't explicitly publish this information, it can be difficult to provide in-house calibration and maintenance support. Firmware or software updates may include changes to MTE specification limits or other performance parameters that are inconsistent with the existing application requirements.

Some manufacturers may not notify MTE users of important firmware or software updates in a timely manner. Therefore, MTE users must take on the task of inquiring about firmware or software upgrades or revisions. Furthermore, all firmware or software updates or version changes must be carefully documented and controlled to ensure that they are compatible with the MTE hardware configuration. The management of firmware and software upgrades or versions must be an integral part of the overall configuration control process.

12.5   Metrology and Calibration Support

The goal of metrology and calibration support is to establish and maintain MTE compliance with performance specifications. As previously discussed, there is no substitute for a complete periodic calibration using external, independent reference standards to determine if MTE parameters or attributes are in conformance or non-conformance with specified tolerance limits.

Firmware or software updates may require the modification of calibration procedures and the adjustment of recall intervals. Built-in calibration factors, tables or curve fit equations for a specific, serial-numbered device must also be periodically updated.

The ability of end user or third party calibration laboratories to calibrate and support firmware-based MTE may be limited. Some complex instruments must be shipped back to the manufacturer for re-calibration and servicing. In these instances, the calibration intervals will primarily be manufacturer driven. Independent verification of MTE firmware or software may also be difficult to achieve, forcing reliance on manufacturer certification.

CHAPTER 13: MEASUREMENT QUALITY AND SPECIFICATIONS

When a measurement supports a decision, the validity and accuracy of the measurement carries the same importance as the decision. Measurement quality assurance (MQA) provides a means for assessing whether or not activities, equipment, environments, and procedures involved in making a measurement produce a result that can be rigorously evaluated for validity and accuracy.

Measurement quality can be evaluated through the application of measurement uncertainty analysis, measurement decision risk analysis and calibration interval analysis. As discussed in Chapter 11, MTE specifications play an important role in all of these analysis methods.

MQA best practices must be an integral part of all activities that impact measurement accuracy and reliability including: defining the measurement requirements, designing measurement systems, selecting commercial off-the-shelf (COTS) MTE, and calibrating and maintaining MTE.

13.1 Defining Measurement Requirements

To assure satisfactory MTE performance, the measurement accuracy and reliability requirements must be clearly defined and documented. To do this, the following questions must be answered:

- What will be measured?

- Why is the measurement being made?

- What decisions will be made from the measurement?

- What measurement confidence limits (i.e., tolerance limits) are required for the decision to be made?

- What level of confidence is needed to assure that the risks of using inadequate measurement data are under control?

- What calibration and maintenance support will the measuring equipment need?

The first step is to identify the physical quantity to be measured. The physical quantity can be measured directly or derived from the measurement of other quantities. If the quantity is derived from other directly measured quantities, then the requirements for each of these measurements must be specified. The functional relationship between the quantity of interest and the measured quantities must also be established.

At a minimum, the following information should be established where applicable for each measured quantity:

1. The static and dynamic characteristics of the quantity or quantities to be measured.

2. The range of measured values.

3. The rate of change, period or frequency of the measured quantity.

4. The measurement method or approach that will be used.

5. The environmental conditions in which the measurement will be made.

The range of values to be measured will be used in the selection of MTE and in the establishment of the full scale (FS) requirements of the designed measurement system. Time and frequency requirements are important in the measurement of transient, periodic, and randomly changing quantities. The rate at which the measured quantity changes and the systematic or repetitive nature of occurrence also affect how the measurement should be made, in selecting the MTE, and in establishing data acquisition requirements.

The confidence limits and confidence level requirements should be established to ensure that the measurement process will meet its quality objective. These requirements should be appropriate for the decision that will be made or actions that will be taken as a result of the measurement. These decision or actions may include one or more of the following:

- To continue or stop a manufacturing process
- To accept or reject a product
- To modify or complete a design
- To take corrective action or withhold it
- To establish scientific fact.

Two important requirements for achieving measurement quality are:

1. The measurement must be traceable.

2. The measurement must have a realistic estimate of its uncertainty.

Measurement traceability requirements are discussed in Chapter 10. A realistic uncertainty estimate means that every element of the measurement process that contributes to uncertainty must be included. The total uncertainty for a measured quantity is comprised of uncertainties due to measurement equipment (e.g., bias and resolution error), repeatability or random error caused by fluctuations in environmental or other ancillary conditions, operator error, etc.[65]

As discussed in Chapter 5, the combined distribution for three or more error sources generally takes on a normal or Gaussian shape, regardless of the shape of the individual error distributions. The standard deviation of the combined error distribution is equal to the total measurement uncertainty. The confidence limits, $\pm L_x$, for the measured quantity, x, can be established using the t-statistic.

$$\pm L_x = \pm t_{\alpha/2, v} \times u_x \qquad (13\text{-}1)$$

where

$$
\begin{aligned}
t_{\alpha/2, v} &= t\text{-statistic} \\
\alpha &= \text{significance level} = 1 - C/100 \\
C &= \text{confidence level (\%)} \\
u_x &= \text{measurement uncertainty} \\
v &= \text{degrees of freedom for } u_x
\end{aligned}
$$

---

[65] Detailed coverage of measurement uncertainty analysis methods and procedures is provided in NASA *Measurement Quality Assurance Handbook* Annex 3 – *Measurement Uncertainty Analysis Principles and Methods*.

As shown in Figure 13-1, $\pm L_x$ represents the confidence or containment limits for values of x. The associated confidence level or containment probability is the area under the distribution curve between these limits.

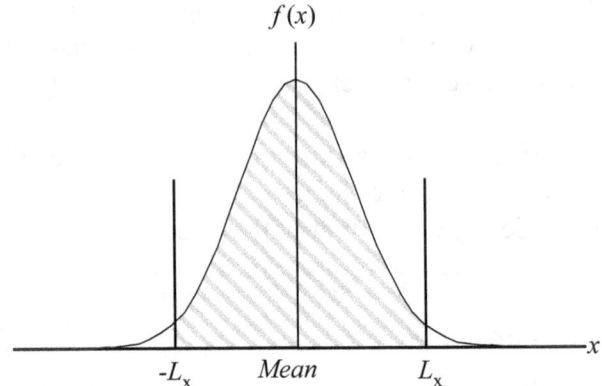

Figure 13-1 Probability Distribution for Measured Quantity x

If the confidence limits are considered to be too large for the intended measurement application or decision making process, then the relative contributions of the measurement process uncertainties to the total uncertainty should be evaluated. As illustrated in Chapter 11, this can be accomplished using a Pareto chart. Once identified, the largest contributors to the total uncertainty can be evaluated for possible mitigation.

When completed, the measurement requirements should specify the following:

- Measured quantity or quantities (voltage, pressure, temperature, etc.)
- Values and range (3 to 10 VDC, 120 to 150 MPa, $-$ 50 to 75 °C, etc.)
- Measurement environment (temperature, pressure, humidity, electromagnetic interference, etc.)
- Frequency range (18 to 20 kHz)
- Confidence limits ($\pm$ 0.1% FS, $\pm$ 0.05 °C, etc.)
- Confidence level (95%, 99.73%, etc.)
- Time period for which the confidence limits apply at the given confidence level (6 months, 5,000 cycles, etc.)

The percentage of measurement data that can be expected to be within the confidence limits at the end of the "guaranteed" time is the end-of-period (EOP) in-tolerance probability or the measurement reliability requirement. The time within which the confidence limits can be "guaranteed" is equivalent to an MTE calibration interval.

13.2 Measurement System Design

The previous section described the development of measurement requirements. This section discusses a structured approach for designing measurement systems to meet these requirements. In this approach, the measurement requirements are translated into performance specifications and then into system design and fabrication.

Figure 13-2 shows an overview of the measurement system development process that includes two design reviews at key stages of system development. The first review occurs after the requirements definition phase and the second review occurs at the completion of the system design phase. The preliminary design review critiques the requirements to establish completeness (e.g., does it meet the required measurement range, bandwidth, etc.). During the critical design review, the system drawings, component parameters and specifications are evaluated to ensure each measurement requirement has been satisfied.

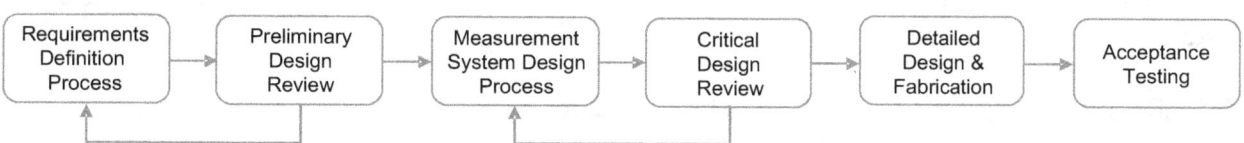

Figure 13-2 Overview of Measurement System Development Process

An example of the measurement system design process is illustrated in Figure 13-3. Both uncertainty and bandwidth considerations are driven by measurement requirements. It is assumed the measurement requirements have been analyzed to establish measurement system specifications and the measurement requirements have been formalized.[66] Once the specifications have been established, it is the designer's responsibility to prove that the system when built will comply with the requirements.

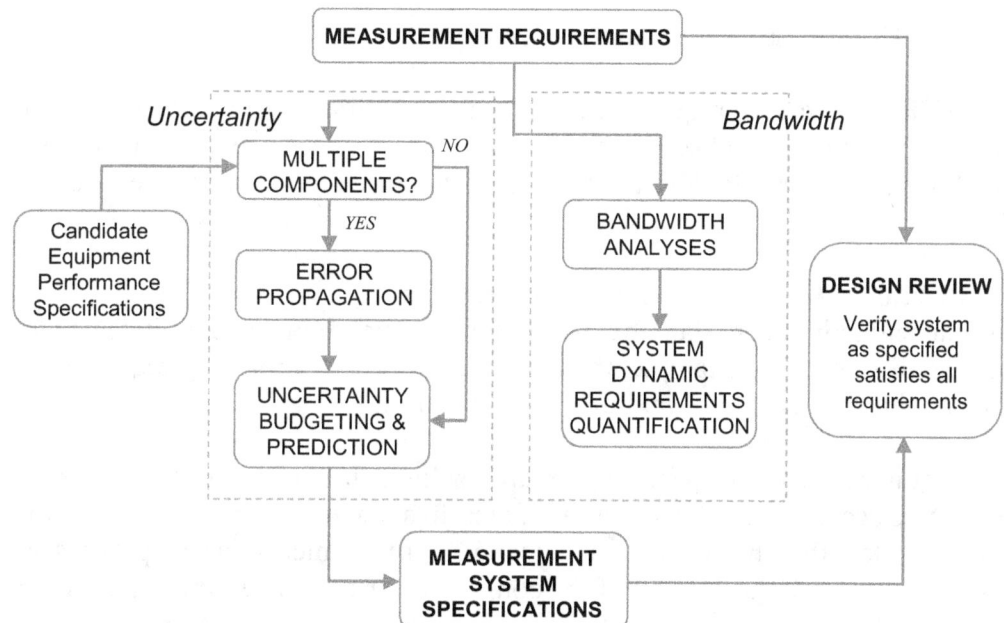

Figure 13-3 Measurement System Design Process

The static and/or dynamic measurement process characteristics are used in the selection of candidate sensors and other system components. Candidate equipment specifications are then used to identify error sources and compute the uncertainty in the output of each module. These uncertainties are propagated from one system module to the next. The overall system output uncertainty is then used to establish the specification limits. An example analysis of a load measurement system is provided in Appendix D.

---

[66] See Chapter 3 of NASA-HNBK-8739.19 NASA *Measurement Quality Assurance Handbook*.

13.3 Selecting Measuring Equipment

The selection of MTE for a given application can often be the most difficult task in ensuring measurement quality. MTE are often initially selected based on manufacturer specified performance capabilities and affordability. The instrument engineer, metrologist or other technical personnel must carefully consider the measurement requirements, intended operating environment, usage level, and potential installation effects.

As discussed in the previous sections, the specific steps associated with designing a measurement system include the following:

1. Identify the physical phenomena to be measured.
2. Specify the measurement requirements.
3. Select candidate measurement equipment and interpret their specifications.
4. Evaluate the effects of changes in environmental conditions on MTE performance.
5. Construct error models for the system components and estimate uncertainties.
6. Establish measurement system performance (i.e., confidence levels and confidence limits).
7. Conduct acceptance tests to verify system performance.
8. Establish MTE calibration and maintenance requirements.

Establishing MTE performance requirements, selecting candidate MTE and acceptance testing are discussed in the following subsections. Measurement system uncertainty analysis is discussed in Appendix D. MTE calibration and maintenance requirements are discussed in Section 13.4.

13.3.1 Establishing MTE Performance Requirements

MTE performance requirements state the functional and operational capabilities needed for the measurement application. For example, the measurement application may require that the MTE be able to respond to rapid changes in the measured quantity.

Performance requirements often specify upper and/or lower tolerance limits for selected MTE parameters such as accuracy, linearity, stability, etc. In some cases, the MTE performance requirements may state that the "expanded" uncertainty of the measurement process shall not exceed some percentage (10% or 25%) of the confidence or tolerance limits specified for the measured quantity.[67]

As discussed in Chapter 11, MTE specifications play an important role in the estimation of measurement process uncertainty. The relative accuracy criterion is a simplified method of assuring that test or calibration process uncertainties do not negatively affect decisions made from the measurements.

The performance requirements should also establish the acceptable ranges or threshold values for the environmental conditions in which the MTE must operate. Environmental conditions may

---

[67] For example, see NASA NSTS 5300.4(1D-2) *Safety, Reliability, Maintainability and Quality Provisions for Space Shuttle Program*, September 1997.

include humidity, temperature, shock, vibration, electromagnetic interference, etc. MTE requirements may also specify interface, compatibility and/or interchangeability with other equipment.

The preparation of MTE performance requirements should involve an integrated team of subject matter experts, including the MTE user and representatives from all functional areas that will be affected by the MTE requirements and/or will be using the measurement data. The MTE performance requirements should be clear and concise so that there is no doubt about the intended use of the equipment and the quality level to which it must perform. The MTE performance requirements must also be verifiable through analysis and testing.

Selected guidance documents for the preparation of measurement equipment performance requirements are listed below.

- MIL-HDBK-2036, *Department of Defense Handbook – Preparation of Electronic Equipment Specifications*, 1999.

- MIL-PRF-28800F, *Performance Specification – Test Equipment for Use with Electrical and Electronic Equipment, General Specification for*, 1996.

MIL-HDBK-2036 provides guidance for developing electronic equipment requirements. MIL-PRF-28800F provides general requirements for MTE used to test and calibrate electrical and electronic equipment. MIL-HDBK-2036 also provides guidance for the evaluation of COTS MTE for use in military applications.

13.3.2 Selecting Candidate MTE

There are various types and designs of MTE commercially available and hundreds more are introduced every year. It is the job of measurement system designers, procurement personnel and end users to select appropriate MTE for the intended application. The selection process should also include quality assurance and metrology personnel to ensure that the candidate MTE can be effectively maintained and calibrated.

When selecting candidate MTE, the user/purchaser must have a good understanding of the basic operating principles of the device to properly review and evaluate whether or not the manufacturer specifications meet the performance requirements for the intended measurement application. Features and characteristics to consider when evaluating candidate MTE include:

- Measurement Range – is the device capable of measuring a wide range or variation in the value of the measured quantity?

- Functionality – are all of the MTE functions necessary to meet the measurement requirements?

- Operating Conditions – can the MTE operate sufficiently in the intended environmental conditions?

- Performance Parameters – are the MTE performance specifications stated clearly and in a measurable, verifiable manner?

- Confidence Level – what confidence level and time period (one month, 6 months, 2 years) must the MTE specification limits be applicable?

Ideally, the MTE performance should not be influenced by extremes in environmental conditions that it may be exposed to during use. However, the MTE response to the combined effects of temperature, pressure, humidity, and vibration may exceed the response to the measured quantity. Consequently, environmental effects can be a significant contributor to MTE error and overall uncertainty in the measured quantity.

Assessing the usage environment and estimating its effect on MTE performance is an important part of the selection and application process. The environmental operating limitations of the MTE should be specified in the manufacturer product data sheet and user manual. The purchaser/user must understand and evaluate the impact that the magnitude and change in environmental conditions can have on MTE performance. In some cases, the purchaser/user may have to conduct tests to determine the effects of any conditions that are not specified by the manufacturer.

Various performance parameters can significantly affect the uncertainty of the measurement process, depending on the MTE configuration and the application. It is the responsibility of the MTE purchaser/user to determine which specifications are important for the measurement application. Unfortunately, there are no industry standards for reporting manufacturer specifications, so caution must be used when interpreting them.[68]

In general, MTE specifications should be closely examined to assure that all the necessary information is available for evaluation. It is a good practice to review performance specifications for similar equipment from different manufacturers to determine whether the manufacturer has listed all relevant performance parameters for the candidate equipment.

The "fine print" should also be examined closely to determine if there are any caveats regarding the specification limits, such as loading effects, frequency response, interface impedances, line power fluctuations (regulation), distortion effects, etc. Note any omissions and specifications that differ significantly from manufacturer to manufacturer. The MTE manufacturer should then be contacted for further information and clarification.

13.3.3 Acceptance Testing

Purchasers/users of newly acquired MTE should work with metrology and calibration personnel to develop and conduct a comprehensive acceptance testing plan to verify that the device will meet the performance requirements. As discussed in Chapter 10, the acceptance tests should provide a comprehensive evaluation of the static and dynamic performance characteristics of the MTE and verify that

- The device is capable of measuring the desired quantities under the required operating conditions.

- The device meets established design requirements and is free of manufacturing defects.

- The device performs within its stated accuracy and other specification limits.

---

[68] See Chapter 9 for more discussion about interpreting MTE specifications.

If the MTE is part of a measurement system, then the acceptance tests should also verify that the device is compatible with other system components.

The acceptance testing program should include: (1) pass/fail criteria, (2) test plan and procedures, (3) data analysis methodologies, and (4) procedures for reporting the test results.

13.4   Calibration and Maintenance

To preserve measurement quality, provisions for periodic MTE calibration should be included when developing measurement requirements. Periodic calibration of MTE parameters provides a major safeguard in controlling uncertainty growth and reducing the risk of substandard performance during use. MTE parameters are calibrated by comparing them to standard values[69] to determine conformance or non-conformance with manufacturer or other specified tolerance limits. MTE parameter calibrations are often subsets of acceptance tests conducted under less stressful operating conditions.

The calibration program should validate MTE performance throughout its life cycle. The interval or period of time between calibrations may vary for each device and/or parameter depending upon the stability, application and degree of use. MTE calibration intervals are set to control uncertainty growth and associated measurement decision risk. Since MTE calibration can be a significant measurement quality assurance cost driver, it is important to optimize MTE calibration intervals by balancing risk and cost. MTE calibration is discussed in Chapter 10.

Calibration procedures must provide instructional guidance to ensure that the calibration activities are performed in a manner consistent with the MTE application requirements. These procedural documents should include the MTE parameters to be calibrated, the operating range(s) tested, the environmental conditions, and the performance criteria used to establish compliance or non-compliance.

> **Note**: Standards such as ISO/IEC 17025:2005 and ANSI/NCSLI Z540.3-2006 also require measurement traceability, as discussed in Chapter 10.

Some MTE users may not perceive periodic calibration as a value added process because they are unaware that MTE parameters can drift or shift out of tolerance due to

- Extensive use
- Exposure to environmental extremes
- Shock and vibration during transport
- Handling or storage.

Consequently, MTE may be selected and purchased without considering the associated calibration and maintenance requirements. In addition, insufficient communication between MTE users and calibration personnel can result in inadequate calibration procedures and inappropriate calibration schedules.

It is a good practice for MTE users and calibration personnel to work closely together to

---

[69] For example, mass or voltage standards, certified reference materials, or other MTE whose accuracy is traceable to the National Institute of Standards and Technology (NIST).

- Determine the reference standards and other equipment needed to calibrate the MTE parameters.

- Develop new calibration procedures or modify existing ones.

- Identify the in-house or external expertise required to calibrate the MTE.

- Specify any training required to operate, calibrate and/or maintain the MTE.

- Determine if an inventory of replacement parts or spare MTE must be maintained.

Calibration personnel should also become familiar with MTE operation in the usage environment to ensure that appropriate standards, equipment and procedures are employed during calibration.

Maintenance and repair becomes necessary when adjustments are inadequate to bring equipment into operational specifications. After maintenance and repair, the MTE performance parameters should be validated by calibration.

MTE maintenance includes technical activities intended to keep instruments, sensors, signal conditioning equipment and other components in satisfactory working condition. Unlike calibration, MTE maintenance is designed to avoid accelerated wear-out and catastrophic operational failures.

MTE maintenance programs should include procedures that ensure:

- Identification of maintenance requirements.

- Timely maintenance of equipment in keeping with the user's performance requirements.

- Periodic scheduling of inspections to verify the effectiveness of the maintenance program.

- Use of manufacturer warranties or servicing agreements, as applicable.

- Establishment of a technical library of applicable maintenance instructions.

MTE maintenance requirements and instructions are usually provided in manufacturer operating manuals. Other maintenance requirements may be established from data or information collected during MTE calibration and/or repair.

Maintenance intervals should be based on mathematical and statistical correlation of historical manufacturer/model failure data that focus on the types of maintenance conducted, the times between maintenance, the failed components or parts, and the time between failures. The historical data are used to establish mean-time-between-failure (MTBF) reliability targets for the various manufacturer/model populations. These MTBF targets determine the appropriate maintenance intervals.

APPENDIX A – TERMS AND DEFINITIONS

It has been a goal of the authors to use consistent terminology throughout this document, even though the terms and definitions employed are designed to be understood across a broad technology base. Where appropriate, terms and definitions have been taken from internationally recognized standards and guidelines in the fields of testing and calibration.

Term	Definition
a priori value	The value indicated before measurements are taken.
Acceleration Error	The maximum difference, at any measured value within the specified range, between output readings taken with and without the application of a specified shock, vibration or constant acceleration along specified axes.
Accelerated Life Testing (ALT)	An activity during product development in which prototypes or first articles are subjected to stresses (i.e., temperature, vibration) at levels higher than those anticipated during actual use.
Acceleration Sensitivity	See Acceleration Error.
Acceptance Testing	Tests conducted to assure that the MTE meets contracted or design requirements.
Accuracy	The conformity of an indicated value of an MTE parameter with an accepted standard value. Manufacturers of instruments and scientific measuring systems often report accuracy as a combined specification that accounts for linearity, hysteresis and repeatability.
Adjusted Mean	The value of a measurement parameter or error obtained by applying a correction factor to a nominal or mean value.
Average-over-period (AOP) Reliability	The in-tolerance probability for an MTE attribute or parameter averaged over its calibration or test interval. The AOP measurement reliability is often used to represent the in-tolerance probability of an MTE attribute or parameter whose usage demand is random over its test or calibration interval.
Amplifier	A device that accepts a varying input signal and produces an output signal that varies in the same way as the input but has a larger amplitude. The input signal may be a current, a voltage, a mechanical motion, or any other signal; the output signal is usually of the same nature.
Analog to Digital Converter (ADC)	A device that converts an analog signal to a digital representation of finite length.
Analog Signal	A quantity or signal that is continuous in both amplitude and time.
Arithmetic Mean	The sum of a set of values divided by the number of values in the set.
Artifact	A physical object or substance with measurable attributes.
Attenuation	Reduction of signal strength, intensity or value.

Term	Definition
Attribute	A measurable characteristic, feature, or aspect of a device, object or substance.
Average	See Arithmetic Mean.
Bandwidth	The range of frequencies that a device is capable of generating, handling, or accommodating; usually the range in which the response is within 3 dB of the maximum response.
Beginning-of-period (BOP) Reliability	The in-tolerance probability for an MTE attribute or parameter at the start of its calibration or test interval.
Bias	A systematic discrepancy between an indicated, assumed or declared value of a quantity and the quantity's true value. See also Attribute Bias and Operator Bias.
Bias Offset	See Offset from Nominal.
Bias Uncertainty	The uncertainty in the bias of a parameter or artifact. The uncertainty in the bias of an attribute or error source quantified as the standard deviation of the bias probability distribution.
Bit	A single character, 0 or 1, in a binary numeral system (base 2). The bit is the smallest unit of storage currently used in computing.
Calibration	A process in which the value of an MTE attribute or parameter is compared to a corresponding value of a measurement reference, resulting in (1) the determination that the parameter or attribute value is within its associated specification or tolerance limits, (2) a documented correction of the parameter or attribute value, or (3) a physical adjustment of the parameter or attribute value.
Calibration Interval	The scheduled interval of time between successive calibrations of one or more MTE parameter or attribute.
Characteristic	A distinguishing trait, feature or quality.
Combined Error	The error comprised of a combination of two or more error sources.
Combined Uncertainty	The uncertainty in a combined error.
Common Mode Rejection (CMR)	The common mode rejection ratio is often expressed in dB using the following relationship: CMR = 20 log(CMRR).
Common Mode Rejection Ratio (CMRR)	Normally defined for amplifiers as the ratio of the signal gain to the ratio of the normal mode voltage and the common-mode voltage. CMMR = Gain/(NMV/CMV). CMRR is a complex function of frequency with its own magnitude and phase.
Common Mode Voltage (CMV)	A voltage which is common to both input terminals of a device with respect to the output reference (usually ground).
Component Uncertainty	The product of the sensitivity coefficient and the standard uncertainty for an error source.

Term	Definition
Computation Error	The error in a quantity obtained by computation. Computation error can result from machine round-off of values obtained by iteration or from the use of regression models. Sometimes applied to errors in tabulated physical constants.
Computed Mean Value	The average value of a sample of measurements.
Confidence Level	The probability that a set of tolerance or containment limits will contain errors for a given error source.
Confidence Limits	Limits that bound errors for a given error source with a specified probability or confidence level.
Containment Limits	Limits that are specified to contain either an attribute or parameter value, an attribute or parameter bias, or other measurement errors.
Containment Probability	The probability that an attribute or parameter value or errors in the measurement of this value will lie within specified containment limits.
Correlation Analysis	An analysis that determines the extent to which two random variables influence one another. Typically the analysis is based on ordered pairs of values. In the context of measurement uncertainty analysis, the random variables are error sources or error components.
Correlation Coefficient	A measure of the extent to which errors from two sources are linearly related. A function of the covariance between the two errors. Correlation coefficients range from minus one to plus one.
Covariance	The expected value of the product of the deviations of two random variables from their respective means. The covariance of two independent random variables is zero.
Coverage Factor	A multiplier used to express an error limit or expanded uncertainty as a multiple of the standard uncertainty.
Creep	The tendency of a solid material to slowly move or deform permanently under the influence of stresses. The rate of this deformation is a function of the material properties, exposure time, exposure temperature and the applied structural load.
Cross-correlation	The correlation between two error sources for two different components of a multivariate analysis.
Cumulative Distribution Function	A mathematical function whose values $F(x)$ are the probabilities that a random variable assumes a value less than or equal to x. Synonymous with Distribution Function.
Cutoff Frequency	The frequency at which the frequency response is $-3dB$ below its maximum value.
Damping	The restraint of vibratory motion, such as mechanical oscillations, noise, and alternating electric currents, by dissipating energy.

124

Term	Definition
	Other types of damping include viscous, coulomb, electrical resistance, radiation, and magnetic.
Deadband	The range through which the input varies without initiating a response (or indication) from the measuring device.
Degrees of Freedom	A statistical quantity that is related to the amount of information available about an uncertainty estimate. The degrees of freedom signifies how "good" the estimate is and serves as a useful statistic in determining appropriate coverage factors and computing confidence limits and other decision variables.
Detection Limit	See Threshold.
Deviation from Nominal	The difference between an attribute's or parameter's measured or true value and its nominal value.
Digital to Analog Converter (DAC)	A device for converting a digital (usually binary) code to a continuous, analog output.
Digital Signal	A quantity or signal that is represented as a series of discrete coded values.
Direct Measurements	Measurements in which a measuring parameter or attribute X directly measures the value of a subject parameter or attribute Y (i.e., X measures Y). In direct measurements, the value of the quantity of interest is obtained directly by measurement and is not determined by computing its value from the measurement of other variables or quantities.
Display Resolution	The smallest distinguishable difference between indications of a displayed value.
Distribution Function	See Cumulative Distribution Function.
Distribution Variance	The mean square dispersion of a random variable about its mean value. See also Variance.
Drift	An undesired change in output over a period of time that is unrelated to input. Drift can result from aging, temperature effects, sensor contamination, etc.
Drift Rate	The amount of drift per unit time. Drift rate can apply to a variety of artifacts and devices, so the type of drift rate and the usage or boundary conditions for which the drift rate applies should always be specified.
Dynamic Performance Characteristics	Those characteristics of a device that relate its response to variations of the physical input with time.
Dynamic Range	The range of input signals that can be converted to output signals by a measuring device or system component.
Effective Degrees of Freedom	The degrees of freedom for Type B uncertainty estimates or a combined uncertainty estimate.

Term	Definition
End-of-Period (EOP) Reliability	The in-tolerance probability for an MTE attribute or parameter at the end of its calibration or test interval.
Equipment Parameter	An aspect or feature of an instrument, measuring device or artifact. See also Attribute.
Error	The arithmetic difference between a measured or indicated value and the true value.
Error Component	The total error in a measured or assumed value of a component variable in a multivariate measurement. For example, in the determination of the volume of a right circular cylinder, there are two error components: the error in the length measurement and the error in the diameter measurement.
Error Distribution	A probability distribution that describes the relative frequency of occurrence of values of a measurement error.
Error Equation	An expression that defines the total error in the value of a quantity in terms of all relevant process or component errors.
Error Limits	Bounding values that are expected to contain the error from a given source with some specified level of probability or confidence.
Error Model	See Error Equation.
Error Source	A parameter, variable or constant that can contribute error to the determination of the value of a quantity.
Error Source Coefficient	See Sensitivity Coefficient.
Error Source Correlation	See Correlation Analysis.
Error Source Uncertainty	The uncertainty in a given error source.
Estimated True Value	The value of a quantity obtained by Bayesian analysis.
Excitation	An external power supply required by measuring devices to convert a physical input to an electrical output. Typically, a well-regulated dc voltage or current.
Expanded Uncertainty	A multiple of the combined standard uncertainty reflecting either a specified confidence level or coverage factor.
False Accept Risk (FAR)	The probability that an equipment attribute or parameter, accepted by conformance testing, will be out-or-tolerance. See NASA-HNBK-8739.19-4 for alternative definitions and applications.
False Reject Risk (FRR)	The probability of an attribute or parameter being in-tolerance and rejected by conformance testing as being out-of-tolerance.
Feature	An MTE attribute that describes enhancements or special characteristics. If the feature reflects a measurable parameter, then it should have a specification that describes its performance.

Term	Definition
Filter	A device that limits the signal bandwidth to reduce noise and other errors associated with sampling.
Frequency Response	The change with frequency of the output/input amplitude ratio (and of phase difference between output and input), for a sinusoidally varying input applied to a measuring device within a stated range of input frequencies.
Full Scale Input (FSI)	The arithmetic difference between the specified upper and lower input limits of a sensor, transducer or other measuring device.
Full Scale Output (FSO)	The arithmetic difference between the specified upper and lower output limits of a sensor, transducer or other measuring device.
Functional Test	A test of a product or device that measures performance parameters to determine whether they meet specifications. Also referred to as a Performance Test.
Gain	The ratio of the output signal to the input signal of an amplifier.
Gain Error	The degree to which gain varies from the ideal or target gain, specified in percent of reading.
Guardband	A supplement specification limit used to reduce the risk of falsely accepting a nonconforming or out-of-compliance MTE parameter.
Highly Accelerated Life Test (HALT)	A process that subjects a product or device to varied accelerated stresses to identify design flaws and establish the stress limits of the product.
Highly Accelerated Stress Screening (HASS)	A post-production testing process that subjects a product or device to various stress levels to identify manufacturing problems. The stress levels employed are typically much lower than those used in HALT.
Heuristic Estimate	An estimate resulting from accumulated experience and/or technical knowledge concerning the uncertainty of an error source.
Hysteresis	The lagging of an effect behind its cause, as when the change in magnetism of a body lags behind changes in an applied magnetic field.
Hysteresis Error	The maximum separation due to hysteresis between upscale-going and downscale-going indications of a measured value taken after transients have decayed.
Independent Error Sources	Two error sources that are statistically independent. See also Statistical Independence.
Instrument	A device for measuring or producing the value of an observable quantity.
In-tolerance	In conformance with specified tolerance limits.

Term	Definition
In-tolerance Probability	The probability that an MTE attribute or parameter value or the error in the value is contained within its specified tolerance limits at the time of measurement.
Least Significant Bit (LSB)	The smallest analog signal value that can be represented with an *n*-bit code. LSB is defined as $A/2^n$, where A is the amplitude of the analog signal.
Level of Confidence	See Confidence Level.
Linearity	A characteristic that describes how a device's output over its range differs from a specified linear response.
Mean Value	*Sample Mean*: The average value of a measurement sample. *Population Mean*: The expectation value for measurements sampled from a population.
Measurement Error	The difference between the measured value of a quantity and its true value.
Measurement Process Errors	Errors resulting from the measurement process (e.g., measurement reference bias, repeatability, resolution error, operator bias, environmental factors, etc).
Measurement Process Uncertainties	The uncertainty in a measurement process error. The standard deviation of the probability distribution of a measurement process error.
Measurement Reference	See Reference Standard
Measurement Reliability	The probability that an MTE attribute or parameter is in conformance with performance specifications. At the measuring device or instrument level, it is the probablity that all attributes or parameters are in conformance or in-tolerance.
Measurement Traceability	See Traceability.
Measurement Uncertainty	The lack of knowledge of the sign and magnitude of measurement error.
Measurement Units	The units, such as volts, millivolts, etc., in which a measurement or measurement error is expressed.
Measuring Device	See Measuring and Test Equipment.
Measuring and Test Equipment (MTE)	A system or device used to measure the value of a quantity or test for conformance to specifications.
Module Error Sources	Sources of error that accompany the conversion of module input to module output.
Module Input Uncertainty	The uncertainty in a module's input error expressed as the uncertainty in the output of the preceding module.
Module Output Equation	The equation that expresses the output from a module in terms of its input. The equation is characterized by parameters that

Term	Definition
	represent the physical processes that participate in the conversion of module input to module output.
Module Output Uncertainty	The combined uncertainty in the output of a given module of a measurement system.
Multiplexer	A multi-channel device designed to accept input signals from a number of sensors or measuring equipment and share downstream signal conditioning components.
Multivariate Measurements	Measurements in which the subject parameter is a computed quantity based on measurements of two or more attributes or parameters.
Noise	Signals originating from sources other than those intended to be measured. The noise may arise from several sources, can be random or periodic, and often varies in intensity.
Nominal Value	The designated or published value of an artifact, attribute or parameter. It may also sometimes refer to the distribution mode value of an artifact, attribute or parameter.
Nonlinearity	See Linearity.
Normal Mode Voltage	The potential difference that exists between pairs of power (or signal) conductors.
Offset	A non-zero output of a device for a zero input.
Operating Conditions	The environmental conditions, such as pressure, temperature and humidity ranges that the measuring device is rated to operate.
Operator Bias	The systematic error due to the perception or influence of a human operator or other agency.
Output Device	See Readout Device.
Overshoot	The amount of output measured beyond the final steady output value, in response to a step change in the physical input.
Parameter	A characteristic of a device, process or function. See also Equipment Parameter.
Parameter Bias	A systematic deviation of a parameter's nominal or indicated value from its true value.
Population	The total set of possible values for a random variable.
Population Mean	The expectation value of a random variable described by a probability distribution.
Precision	The number of places past the decimal point in which the value of a quantity can be expressed. Although higher precision does not necessarily mean higher accuracy, the lack of precision in a measurement is a source of measurement error.

Term	Definition
Probability	The likelihood of the occurrence of a specific event or value from a population of events or values.
Probability Density Function (pdf)	A mathematical function that describes the relative frequency of occurrence of the values of a random variable.
Quantization	The sub-division of the range of a reading into a finite number of steps, not necessary equal, each of which is assigned a value. Particularly applicable to analog to digital and digital to analog conversion processes.
Quantization Error	Error due to the granularity of resolution in quantizing a sampled signal. Contained within +/- 1/2 LSB (least significant bit) limits.
Random Error	See Repeatability.
Range	An interval of values for which specified tolerances apply. In a calibration or test procedure, a setting or designation for the measurement of a set of specific points.
Rated Output (RO)	See Full Scale Output.
Readout Device	A device that converts a signal to a series of numbers on a digital display, the position of a pointer on a meter scale, tracing on recorder paper or graphic display on a screen.
Reference Standard	An artifact used as a measurement reference whose value and uncertainty have been determined by calibration and documented.
Reliability	The probability that an MTE parameter, component, or part will perform its required function under defined conditions for a specified period of time.
Reliability Model	A mathematical function relating the in-tolerance probability of one or more MTE attributes or parameters and the time between calibration. Used to project uncertainty growth over time.
Repeatability	The error that manifests itself in the variation of the results of successive measurements of a quantity carried out under the same measurement conditions and procedure during a measurement session. Often referred to as Random Error.
Reproducibility	The closeness of the agreement between the results of measurements of the value of a quantity carried out under different measurement conditions. The different conditions may include: principle of measurement, method of measurement, observer, measuring instrument(s), reference standard, location, conditions of use, time.
Resolution	The smallest discernible value indicated by a measuring device.
Resolution Error	The error due to the finiteness of the precision of a measurement.
Response Time	The time required for a sensor output to change from its previous state to a final settled value.

Term	Definition
Sample	A collection of values drawn from a population from which inferences about the population are made.
Sample Mean	The arithmetic average of sampled values.
Sample Size	The number of values that comprise a sample.
Sensitivity	The ratio between a change in the electrical output signal to a small change in the physical input of a sensor or transducer. The derivative of the transfer function with respect to the physical input.
Sensitivity Coefficient	A coefficient that weights the contribution of an error source to a combined error.
Sensor	Any of various devices designed to detect, measure or record physical phenomena.
Settling Time	The time interval between the application of an input and the time when the output is within an acceptable band of the final steady-state value.
Signal Conditioner	A device that provides amplification, filtering, impedance transformation, linearization, analog to digital conversion, digital to analog conversion, excitation or other signal modification.
Span	See Dynamic Range.
Specification	A numerical value or range of values that bound the performance of an MTE parameter or attribute.
Stability	The ability of a measuring device to give constant output for a constant input over a period of time under specified environmental and/or ancillary conditions.
Standard Deviation	The square root of the variance of a sample or population of values. A quantity that represents the spread of values about a mean value. In statistics, the second moment of a distribution.
Standard Uncertainty	The standard deviation of an error distribution.
Static Performance Characteristic	An indication of how the measuring equipment or device responds to a steady-state input at one particular time.
Statistical Independence	A property of two or more random variables such that their joint probability density function is the product of their individual probability density functions. Two error sources are statistically independent if one does not exert an influence on the other or if both are not consistently influenced by a common agency.
Student's t-statistic	Typically expressed as $t_{\alpha,\nu}$, it denotes the value for which the distribution function for a t-distribution with ν degrees of freedom is equal to $1 - \alpha$. A multiplier used to express an error limit or expanded uncertainty as a multiple of the standard uncertainty.

131

Term	Definition
Symmetric Distribution	A probability distribution of random variables that are equally likely to be found above or below a mean value.
System Equation	A mathematical expression that defines the value of a quantity in terms of its constituent variables or components.
System Module	An intermediate stage of a system that transforms an input quantity into an output quantity according to a module output equation.
System Output Uncertainty	The total uncertainty in the output of a measurement system.
t Distribution	A symmetric, continuous distribution characterized by the degrees of freedom parameter. Used to compute confidence limits for normally distributed variables whose estimated standard deviation is based on a finite degrees of freedom. Also referred to as the Student's t-distribution.
Temperature Coefficient	A quantitative measure of the effects of a variation in operating temperature on a device's zero offset and sensitivity.
Temperature Effects	The effect of temperature on the sensitivity and zero output of a measuring device.
Thermal Drift	The change in output of a measuring device per degree of temperature change, when all other operating conditions are held constant.
Thermal Sensitivity Shift	The variation in the sensitivity of a measuring device as a function of temperature.
Thermal Transient Response	A change in the output from a measuring device generated by temperature change.
Thermal Zero Shift	The shift in the zero output of a measuring device due to change in temperature.
Threshold	The smallest change in the physical input that will result in a measurable change in transducer output.
Time Constant	The time required to complete 63.2% of the total rise or decay after a step change of input. It is derived from the exponential response $e^{-t/\tau}$ where t is time and τ is the time constant.
Tolerance Limits	Typically, engineering tolerances that define the maximum and minimum values for a product to work correctly. These tolerances bound a region that contains a certain proportion of the total population with a specified probability or confidence.
Total Module Uncertainty	See Module Output Uncertainty.
Total Uncertainty	The standard deviation of the probability distribution of the total combined error in the value of a quantity obtained by measurement.
Total System Uncertainty	See System Output Uncertainty.

Term	Definition
Traceability	The property of the result of a measurement or the value of a standard whereby it can be related to stated references, usually national or international standards, through an unbroken chain of comparisons, all having stated uncertainties.
Transducer	A device that converts an input signal of one form into an output signal of another form.
Transfer Function	A mathematical equation that shows the functional relationship between the physical input signal and the electrical output signal.
Transient Response	The response of a measuring device to a step-change in the physical input. See also Response Time and Time Constant.
Transverse Sensitivity	An output caused by motion, which is not in the same axis that the device is designed to measure. Defined in terms of output for cross-axis input along the orthogonal axes.
True Value	The value that would be obtained by a perfect measurement. True values are by nature indeterminate.
Uncertainty	See Standard Uncertainty.
Uncertainty Component	The uncertainty in an error component.
Uncertainty in the Mean Value	The standard deviation of the distribution of mean values obtained from multiple sample sets for a given measured quantity. Estimated by the standard deviation of a single sample set divided by the square root of the sample size.
Uncertainty Growth	The increase in the uncertainty in the value of a parameter or attribute over the time elapsed since measurement.
Unit Under Test (UUT)	An MTE submitted for test or calibration.
Validation	Proof that an MTE accomplishes the intended purpose. Validation may be determined by a combination of test and demonstration.
Variance	(1) Population: The expectation value for the square of the difference between the value of a variable and the population mean. (2) Sample: A measure of the spread of a sample equal to the sum of the squared observed deviations from the sample mean divided by the degrees of freedom for the sample. Also referred to as the mean square error.
Verification	The set of operations that assures that specified requirements have been met. (Fluke Calibration: Philosophy in Practice) Proof of compliance with performance specifications. Verification may be determined by testing, analysis, demonstration, inspection or a combination thereof. (NASA Systems Engineering Handbook)
Vibration Sensitivity	The maximum change in output, at any physical input value within the specified range, when vibration levels of specified amplitude and range of frequencies are applied to measuring device along specified axes.

Term	Definition
Warm-up Time	The time it takes a circuit to stabilize after the application of power.
Within Sample Sigma	An indicator of the variation within samples.
Zero Balance	See Offset.
Zero Drift	See Zero Shift.
Zero Offset	See Offset.
Zero Shift	A change in the output of a measuring device, for a zero input, over a specified period of time.

APPENDIX B - STANDARDS FOR THE TESTING AND REPORTING OF MTE PERFORMANCE CHARACTERISTICS

Table B-1. Testing Standards for Transducers

Document	MTE Type
ISA-37.10-1982 (R1995) *Specifications and Tests for Piezoelectric Pressure and Sound Pressure Transducers*	Piezoelectric Pressure and Sound Pressure Transducers
ISA-37.3-1982 (R1995) *Specifications and Tests for Strain Gage Pressure Transducers*	Strain Gage Pressure Transducers
ISA-37.6-1982 - (R1995) *Specifications and Tests for Potentiometric Pressure Transducers*	Potentiometric Pressure Transducers
ISA-37.16.01-2002 *A Guide for the Dynamic Calibration of Pressure Transducers*	Pressure Transducers
ASME B40.100 - 2005 *Pressure Gauges and Gauge Attachments*	Pressure Gauges
ISA-37.8-1982 (R1995) *Specifications and Tests for Strain Gage Force Transducers*	Strain Gage Force Transducers
ASTM E74 - 06 *Standard Practice of Calibration of Force-Measuring Instruments for Verifying the Force Indication of Testing Machines*	Force Measuring Instruments
ISA-37.12-1982 (R1995) *Specifications and Tests for Potentiometric Displacement Transducers*	Potentiometric Displacement Transducers
ISA-MC96.1-1982 *Temperature Measurement Thermocouples*	Thermocouples
ISA-RP37.2-1982 (R1995) *Guide for Specifications and Tests for Piezoelectric Acceleration Transducers for Aerospace Testing*	Piezoelectric Acceleration Transducers
ISA-37.5-1982 - (R1995) *Specifications and Tests for Strain Gage Linear Acceleration Transducers*	Strain Gage Acceleration Transducers
ISA-RP31.1-1977 *Specification, Installation and Calibration of Turbine Flowmeters*	Turbine Flowmeters
ASME MFC-10M - 2000 *Method for Establishing Installation Effects on Flow Meters*	Flow Meters
ASTM D3195-90(2004) *Standard Practice for Rotameter Calibration*	Flow Meter

Document	MTE Type
ANSI S1.15-2005/Part 2 *Measurement Microphones - Part 2: Primary Method for Pressure Calibration of Laboratory Standard Microphones by the Reciprocity Technique*	Microphones
ANSI S1.16-2000 (R2005) *Method for Measuring the Performance of Noise Discriminating and Noise Canceling Microphones*	Microphones
ANSI S1.20-1988 (R2003) *Procedures for Calibration of Underwater Electroacoustic Transducers*	Electroacoustic Transducers
ANSI S1.40-2006 *American National Standard Specifications and Verification Procedures for Sound Calibrators*	Sound Calibrators
ANSI S12.5-2006/ISO 6926:1999 *Acoustics - Requirements for the Performance and Calibration of Reference Sound Sources Used for the Determination of Sound Power Levels*	Acoustic Sound Sources
ANSI S2.2-1959 (R2006) *Methods for the Calibration of Shock and Vibration Pickups*	Electro-mechanical Shock and Vibration Transducers
ISA-26-1968 *Dynamic Response Testing of Process Control Instrumentation.*	Process Control Instrumentation

Table B-2. Testing Standards for Dimensional and Volumetric Measuring Devices

Document	MTE Type
ASME B89.1.5 – 1998 *Measurement of Plain External Diameters for use as Master Discs or Cylindrical Plug Gages* (Reaffirmed 2004)	Plug Gages
ASME B89.1.6 - 2002 *Measurement of Plain Internal Diameters for use as a Master Rings or Ring Gauges*	Ring Gauges
ASME B89.1.9 - 2002 *Gage Blocks*	Gage Blocks
ASME B89.1.2M - 1991 *Calibration of Gage Blocks by Contact Comparison Methods, Through 20 in. and 500 mm*	Gage Blocks
ASME B89.1.10M - 2001 *Dial Indicators (for Linear Measurement)*	Dial Indicators
ASME B89.1.13 - 2001 *Micrometers*	Micrometers
ASME B89.1.17 - 2001 *Measurement of Thread Measuring Wires*	Thread Measuring Wires
ASME B89.6.2 - 1973 *Temperature and Humidity Environment for Dimensional Measurement (Reaffirmed 2003)*	Dimensional
ASME B89.4.22 - 2004 *Methods for Performance Evaluation of Articulated Arm Coordinate Measuring Machines (CMM)*	Coordinate Measuring Machines
ASME B89.4.19 - 2006 *Performance Evaluation of Laser-Based Spherical Coordinate Measurement Systems*	Coordinate Measuring Machines
ASME B89.4.10360.2-2008 *Acceptance Test and Reverification Test for Coordinate Measuring Machines (CMMs) Part 2: CMMs Used for Measuring Linear Dimensions*	Coordinate Measuring Machines
ANSI/ASME B89.4.1-1997 *Methods for Performance Evaluation of Coordinate Measuring Machines*	Coordinate Measuring Machines
ASME B89 - 1990 *Technical Report 1990, Parametric Calibration of Coordinate Measuring Machines (CMM)*	Coordinate Measuring Machines
ASTM E1157 – 87(2006) *Standard Specification for Sampling and Testing of Reusable Laboratory Glassware*	Graduated and Ungraduated Glassware
ASTM E542 – 01(2007) *Standard Practice for Calibration of Laboratory Volumetric Apparatus*	Graduated Cylinders, Flasks and Pipettes

Table B-3. Testing Standards for Complex Instruments

Document	MTE Type
ASTM E10 – 07A *Standard Test Method for Brinell Hardness of Metallic Materials*	Brinell Hardness
ASTM E18 – 08A *Standard Test Methods for Rockwell Hardness of Metallic Materials*	Rockwell Hardness
ASTM E251 – 92(2003) *Standard Test Methods for Performance Characteristics of Metallic Bonded Resistance Strain Gages*	Strain Gages
ASTM E766 – 98(2008) *Standard Practice for Calibrating the Magnification of a Scanning Electron Microscope*	Scanning Electron Microscope
ASTM E915 – 96(2002) *Standard Test Method for Verifying the Alignment of X-Ray Diffraction Instrumentation for Residual Stress Measurement*	X-Ray Diffraction
ASTM E986 – 04 *Standard Practice for Scanning Electron Microscope Beam Size Characterization*	Scanning Electron Microscope
ASTM E2428 – 08 *Standard Practice for Calibration of Torque-Measuring Instruments for Verifying the Torque Indication of Torque Testing Machines*	Torque
ASTM E275 – 01 *Standard Practice for Describing and Measuring Performance of Ultraviolet, Visible, and Near-Infrared Spectrophotometers*	Spectrophotometers
ASTM E388 – 04 *Standard Test Method for Wavelength Accuracy of Spectral Bandwidth Fluorescence Spectrometers*	Fluorescence Spectrometers
ASTM E516 – 95A (2005) *Standard Practice for Testing Thermal Conductivity Detectors Used in Gas Chromatography*	Gas Chromatographs
ASTM E578 – 07 *Standard Test Method for Linearity of Fluorescence Measuring Systems*	Fluorescence Measuring Systems
ASTM E594 – 96 (2006) *Standard Practice for Testing Flame Ionization Detectors Used in Gas or Supercritical Fluid Chromatography*	Gas Chromatograph
ASTM E685 – 93 (2005) *Standard Practice for Testing Fixed-Wavelength Photometric Detectors Used in Liquid Chromatography*	Liquid Chromatograph
ASTM E932 – 89 (2007) *Standard Practice for Describing and Measuring Performance of Dispersive Infrared Spectrometers*	Infrared Spectrometer

APPENDIX C - GENERALIZED TRANSFER FUNCTIONS FOR SELECTED MTE

The equations listed in Table C-1 are general transfer functions that include basic error terms. These equations are not considered to be definitive. The actual form of the equations will vary depending on the MTE specifications. The modification of the load cell transfer function using manufacturer specifications is illustrated in Section 9.2.

Table C-1. General Transfer Functions for Selected MTE

Transfer Function	Error Sources
Sensor or Transducer (No Excitation) $Y_{out} = X_{in} \times S + Y_{os} + Y_e$ Y_{out} = electrical output X_{in} = physical input quantity S = sensitivity Y_{os} = zero offset Y_e = sensor or transducer error	• Reference junction • Nonlinearity • Hysteresis • Noise • Resolution • Repeatability • Offset error • Temperature effects • Long-term stability
Sensor or Transducer (Excitation) $Y_{out} = X_{in} \times S \times E + Y_{os} + Y_e$ Y_{out} = electrical output X_{in} = physical input quantity S = sensitivity E = excitation voltage or current Y_{os} = zero offset Y_e = sensor or transducer error	• Nonlinearity • Hysteresis • Noise • Resolution • Repeatability • Offset error • Temperature effects • Long-term stability
Excitation or External Power Source $E = E_n + E_e$ E = excitation voltage or current E_n = nominal excitation voltage or current E_e = excitation error	• Accuracy • Drift • Stability
Precision Input or Reference Standard $RS = RS_n + RS_e$ RS = reference standard value RS_n = nominal or indicated value RS_e = reference standard error	• Bias • Drift • Stability • Environmental
Simulator or Calibrator $Y_{out} = Y_n + Y_{os} + Y_e$ Y_{out} = indicated output value Y_n = simulator or calibrator output Y_{os} = zero offset Y_e = simulator or calibrator error	• Hysteresis • Noise • Resolution • Repeatability • Offset error • Temperature effects • Stability

Transfer Function	Error Sources
Analog or Digital Multiplexer $Y_{out} = X_{in} + Y_e$ Y_{out} = electrical output X_{in} = analog or digital input Y_e = multiplexer error	• Cross-talk • Thermal-induced Voltages
Amplifier $Y_{out} = X_{in} \times G + Y_e$ Y_{out} = output X_{in} = input G = gain Y_e = amplifier error	• Gain accuracy • Gain stability • CMMR • Noise • Nonlinearity • Offset error • Settling Time • Slew Rate • Overshoot • Temperature effects
Charge Amplifier $Y_{out} = -Q_{in} / C_r + Y_e$ Y_{out} = voltage output Q_{in} = current input C_r = range capacitor Y_e = electrical input	• Drift due to leakage current • Drift due to offset voltage
Filter $Y_{out} = X_{in} \times A + Y_e$ Y_{out} = output X_{in} = input A = attenuation Y_e = filtering error	• Linearity • Offset error • Thermal effects • Stability • Passband Ripple & Flatness • Stopband Ripple & Flatness • Stopband Attenuation
AD and DA Converters $Y_{out} = X_{in} + Y_e$ Y_{out} = digital or analog output X_{in} = analog or digital input Y_e = ADC or DAC error	• Aperture error • Sampling distortion • Stabilization • Quantization • Nonlinearity • Offset error • Supply rejection • Temperature effects • Long-term drift
Cables and Connectors $Y_{out} = X_{in} \times A + Y_e$ Y_{out} = electrical output X_{in} = electrical input A = attenuation Y_e = cabling or connector error	• Attenuation • Loading • Interface Errors
Data Acquisition $Y_{out} = Y_{in} + Y_e$ Y_{in} = input Y_{out} = output Y_e = DAQ error	• Time Code Error • Signal Conditioning Error

Transfer Function	Error Sources
Data Processors $Y_{out} = Y_{in} + Y_e$ Y_{in} = input Y_{out} = output Y_e = processor error	• Regression Error • Round-off/Truncation Error
Output Displays $Y_{out} = Y_{in} + Y_e$ Y_{out} = output Y_{in} = input Y_e = resolution error	• Resolution

APPENDIX D - LOAD MEASUREMENT SYSTEM ANALYSIS

This analysis considers a force or load measurement with the following requirements:

Measurement range: 0 to 5 lb$_f$
Measurement environment: 75 to 85 °F, 14.7 psi, 10 to 15 % RH
Confidence limits: ± 0.05 lb$_f$ (± 1% of full scale)
Confidence level: 99%
Time period for confidence level: 1 year

As shown in Figure D-1, the load will be measured using a multi-component system consisting of a load cell, amplifier, digital multimeter and data processor. For simplicity, errors due to cables, connectors and other interface components are not considered in this analysis.

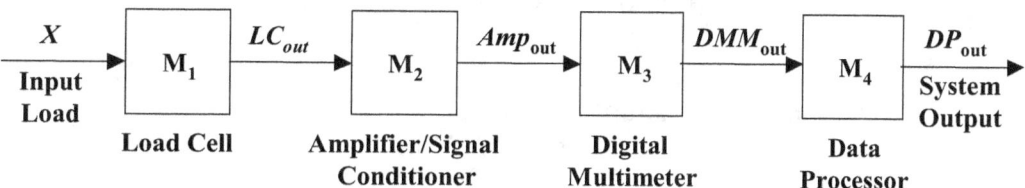

Figure D-1 Block Diagram of Load Measurement System

The output from any given component or module of the system comprises the input of subsequent module. Since each module's output carries with it an element of uncertainty, this means that this uncertainty will propagate through subsequent modules.

After a system diagram has been established, the next step is to develop the equations that relate the inputs and outputs for each module. The basic approach is to describe the physical processes that transform the system input along its path from module to module. In this example, the system input is a 3 lb$_f$ load.

D.1   Load Cell Module (M₁)

The first module in the measurement system consists of an MDB-5-T load cell manufactured by Transducer Techniques, Inc. The module equation and procedure for estimating the uncertainty in the load cell output were developed in Chapter 11. The estimated uncertainties and sensitivity coefficients for each load cell parameter are duplicated below for reference.

Table D-1. Estimated Uncertainties for Load Cell Parameters

Param. Name	Nominal or Stated Value	± Error Limits	Conf. Level	Standard Uncertainty	Sensitivity Coefficient	Component Uncertainty
NL	0 mV/V	± 0.001 mV/V	95	0.0005 mV/V	8 V	0.0041 mV
Hys	0 mV/V	± 0.001 mV/V	95	0.0005 mV/V	8 V	0.0041 mV
NS	0 mV/V	± 0.001 mV/V	95	0.0005 mV/V	8 V	0.0041 mV
ZO	0 mV/V	± 0.02 mV/V	95	0.0102 mV/V	8 V	0.0816 mV
$TR_{°F}$	10 °F	± 2.0 °F	99	0.7764 °F	0	
TE_{out}	0 lb$_f$/°F	± 1.5 × 10^{-4} lb$_f$/°F	95	7.65 × 10^{-5} lb$_f$/°F	32 °F × mV/lb$_f$	0.0024 mV
TE_{zero}	0 mV/°F	± 0.0001 mV/V/°F	95	5 × 10^{-5} mV/V/°F	80 °F × V	0.0041 mV
V_{Ex}	8 V	± 0.25 V	95	0.1276 V	1.2 mV/V	0.1531 mV

The resulting output voltage, uncertainty degrees of freedom for a 3 lb$_f$ input load are summarized below.

$$LC_{out} = 9.6 \text{ mV}, \ u_{LC_{out}} = 0.174 \text{ mV}, \ v_{LC_{Out}} = \infty$$

D.2 Amplifier Module (M$_2$)

The second system module is a TMO-2 Amplifier/Signal Conditioner, manufactured by Transducer Techniques Inc. This module amplifies the mV output from the load cell module to V. This module also supplies 8 VDC ± 0.25 V to the load cell. The nominal amplifier gain is the ratio of the maximum amplifier output to the maximum load cell output. The basic transfer function for this module is given in equation (D-1).

$$Amp_{out} = LC_{out} \times G \qquad\qquad (D\text{-}1)$$

where

$$\begin{aligned} Amp_{out} &= \text{Amplifier Output, V} \\ G &= \text{Amplifier Gain, V/mV} \end{aligned}$$

For this module, the following error sources must be considered:

- Load cell error
- Amplifier error

Manufacturer's published specifications for the amplifier[70] are listed in Table D-2. For a recommended applied excitation voltage of 10 VDC, the MDB-5-T load cell has a maximum rated output of 20 mV. Therefore, the TMO-2 amplifier has a nominal gain of 10V/20 mV or 0.5 V/mV.

Table D-2. Specifications for TMO-2 Amplifier

Specification	Value	Units
Maximum Output Voltage	10	V
Gain (nominal)	0.5	V/mV
Gain Accuracy	0.05% of Full Scale	mV
Gain Stability	0.01%	mV
Nonlinearity	0.01%	mV
Noise and Ripple	< 3	mV
Balance Stability	0.2%	mV
Temperature Coefficient	0.02% of F.S./°C	mV/°C

Given the above specifications, the following sources of amplifier error are applicable to this analysis:

- Gain accuracy
- Gain stability (or Instability)
- Nonlinearity
- Noise

---

[70] Specifications obtained from www.ttloadcells.com/TMO-2.cfm

- Balance stability
- Temperature coefficient

Gain Accuracy
Gain is the ratio of the amplifier output signal voltage to the input signal voltage. In this case, the TMO-2 amplifier has a nominal gain of 10V/20 mV or 0.5 V/mV. The manufacturer specified accuracy of ± 0.05% of full scale is interpreted to be a 95% confidence interval for normally distributed errors.

Gain Stability
If the amplifier voltage gain is represented by G_V, its input resistance by R and its feedback resistance by R_f, then oscillations are possible when

$$\frac{RG_V}{R + R_f} = \pi .$$

These oscillations appear as instability in the amplifier gain. The manufacturer specification of 0.01% is interpreted to be ± 0.01% of full scale. These limits are assumed to represent a 95% confidence interval for normally distributed errors.

Nonlinearity
As with the load cell module, actual amplifier response may depart from the ideal or assumed output versus input curve. Nonlinearity errors are point-by-point differences in actual versus expected response over the range of input signal levels. The manufacturer specification of 0.01% is interpreted to be ± 0.01% of full scale and representative of a 95% confidence interval for normally distributed errors.

Noise
Noise generated within the amplifier, that enters the signal path, causes errors in the amplifier output. Since noise is directly related to gain, manufacturers usually specify noise error in absolute units of Volts RMS or Volts peak-to-peak. The manufacturer specification of 3 mV peak-to-peak is estimated to be ± 1.5 mV limits that are equivalent to a 99% confidence interval for normally distributed errors.

Balance Stability
Balance stability, or instability, refers to a non-zero amplifier output exhibited for a zero input. Although balance instability can be reduced by adjustment, there is no way to completely eliminate it because we do not know the true value of the zero offset. The manufacturer specification of ± 0.2% is interpreted to be ± 0.2% of full scale. These limits are also interpreted to be a 95% confidence interval for normally distributed errors.

Temperature Coefficient
Both the balance (zero) and gain are affected by temperature. Manufacturers generally state this as a temperature coefficient in terms of percent change or full scale per degree. The manufacturer specification of ± 0.02% of full scale/°C is interpreted to be a 95% confidence interval for normally distributed errors.

To quantify the effect of temperature, however, we must establish the expected temperature change and use this with the temperature coefficient to compute expected variations. As with the load cell module, the impact of temperature correction error is estimated using a temperature range of 10 °F (5.6 °C) with measurement error limits of ± 1.1 °C with an associated confidence level of 99% for normally distributed errors.

The appropriate transfer function or output equation for the amplifier module is given in equation (D-2). The parameters used in the amplifier output equation are listed in Table D-3. The normal distribution is applied for all parameters.

$$Amp_{out} = LC_{out} \times G + G_{Acc} + G_S + G_{NL} + G_{NS} + B_{St} + TC \times TR_{\circ C} \qquad \text{(D-2)}$$

Table D-3. Parameters used in Amplifier Module Equation

Parameter Name	Description	Nominal or Mean Value	Error Limits	Percent Confidence
LC_{out}	Amplifier Input			
G	Gain	0.5 V/mV		
G_{Acc}	Gain Accuracy	0 V	± 5 mV	95
G_S	Gain Stability	0 V	± 1 mV	95
G_{NL}	Nonlinearity	0 V	± 1 mV	95
G_{NS}	Noise	0 V	± 1.5 mV	99
B_{St}	Balance Stability	0 V	± 20 mV	95
TC	Temperature Coefficient	0 V/°C	± 2 mV/°C	95
$TR_{\circ C}$	Temperature Range	5.6 °C	± 1.1 °C	99

From equation (D-2), the error model for the amplifier module is given in equation (D-3).

$$\varepsilon_{Amp_{out}} = c_{LC_{out}} \varepsilon_{LC_{out}} + c_G \varepsilon_G + c_{G_{Acc}} \varepsilon_{G_{Acc}} + c_{G_S} \varepsilon_{G_S} + c_{G_{NL}} \varepsilon_{G_{NL}}$$
$$+ c_{G_{NS}} \varepsilon_{G_{NS}} + c_{B_{St}} \varepsilon_{B_{St}} + c_{TC} \varepsilon_{TC} + c_{TR_{\circ C}} \varepsilon_{TR_{\circ C}} \qquad \text{(D-3)}$$

The partial derivative equations used to compute the sensitivity coefficients are listed below.

$$c_{LC_{out}} = \frac{\partial Amp_{out}}{\partial LC_{out}} = G \qquad c_G = \frac{\partial Amp_{out}}{\partial G} = LC_{out} \qquad c_{G_{Acc}} = \frac{\partial Amp_{out}}{\partial G_{Acc}} = 1$$

$$c_{G_S} = \frac{\partial Amp_{out}}{\partial G_S} = 1 \qquad c_{G_{NL}} = \frac{\partial Amp_{out}}{\partial G_{NL}} = 1 \qquad c_{G_{NS}} = \frac{\partial Amp_{out}}{\partial G_{NS}} = 1$$

$$c_{B_{St}} = \frac{\partial Amp_{out}}{\partial B_{St}} = 1 \qquad c_{TC} = \frac{\partial Amp_{out}}{\partial TC} = TR_{\circ C} \qquad c_{TR_{\circ C}} = \frac{\partial Amp_{out}}{\partial TR_{\circ C}} = TC$$

The uncertainty model for the amplifier module output is developed by applying the variance operator to the error model.

$$u_{Amp_{out}} = \sqrt{var\left(\varepsilon_{Amp_{out}}\right)}$$

$$= \sqrt{var\left(\begin{array}{c} c_{LC_{out}}\varepsilon_{LC_{out}} + c_G\varepsilon_G + c_{G_{Acc}}\varepsilon_{G_{Acc}} + c_{G_S}\varepsilon_{G_S} + c_{G_{NL}}\varepsilon_{G_{NL}} \\ + c_{G_{NS}}\varepsilon_{G_{NS}} + c_{B_{St}}\varepsilon_{B_{St}} + c_{TC}\varepsilon_{TC} + c_{TR_{\circ C}}\varepsilon_{TR_{\circ C}} \end{array}\right)} \quad \text{(D-4)}$$

There are no correlations between error sources, so the uncertainty model for the amplifier module output can be expressed as

$$u_{Amp_{out}} = \sqrt{\begin{array}{c} c_{LC_{out}}^2 u_{LC_{out}}^2 + c_G^2 u_G^2 + c_{G_{Acc}}^2 u_{G_{Acc}}^2 + c_{G_S}^2 u_{G_S}^2 + c_{G_{NL}}^2 u_{G_{NL}}^2 \\ + c_{G_{NS}}^2 u_{G_{NS}}^2 + c_{B_{St}}^2 u_{B_{St}}^2 + c_{TC}^2 u_{TC}^2 + c_{TR_{\circ C}}^2 u_{TR_{\circ C}}^2 \end{array}} \quad \text{(D-5)}$$

As previously discussed, all of the error sources identified for the amplifier module are assumed to follow a normal distribution. Therefore, the corresponding uncertainties can be estimated from the error limits, confidence level, and the inverse normal distribution function.

$$u = \frac{L}{\Phi^{-1}\left(\dfrac{1+p}{2}\right)} \quad \text{(D-6)}$$

For example, the uncertainty in the gain accuracy is estimated to be

$$u_{G_{Acc}} = \frac{5\,\text{mV}}{\Phi^{-1}\left(\dfrac{1+0.95}{2}\right)} = \frac{5\,\text{mV}}{1.9600} = 2.551\,\text{mV}.$$

The sensitivity coefficients are computed using the parameter nominal values.

$$c_{LC_{out}} = G = 0.5\,\text{V/mV} \qquad c_G = LC_{out} = 9.6\,\text{mV} \qquad c_{G_{Acc}} = 1$$

$$c_{G_S} = 1 \qquad c_{G_{NL}} = 1 \qquad c_{G_{NS}} = 1$$

$$c_{B_{St}} = 1 \qquad c_{TC} = TR_{\circ C} = 5.6\,^\circ\text{C} \qquad c_{TR_{\circ C}} = TC = 0$$

The estimated uncertainties and sensitivity coefficients for each parameter are listed in Table D-4.

Table D-4. Estimated Uncertainties for Amplifier Module Parameters

Param. Name	Nominal or Mean Value	± Error Limits	Percent Confid.	Standard Uncertainty	Sensitivity Coefficient	Component Uncertainty
LC_{out}	9.6 mV			0.1740 mV	0.5 V/mV	0.0869 V
G	0.5 V/mV				9.6 mV	
G_{Acc}	0 V	± 5 mV	95	2.551 mV	1	0.0026 V
G_S	0 V	± 1 mV	95	0.510 mV	1	0.0005 V

Param. Name	Nominal or Mean Value	± Error Limits	Percent Confid.	Standard Uncertainty	Sensitivity Coefficient	Component Uncertainty
G_{NL}	0 V	± 1 mV	95	0.510 mV	1	0.0005 V
G_{NS}	0 V	± 1.5 mV	99	0.583 mV	1	0.0006 V
B_{St}	0 V	± 20 mV	95	10.204 mV	1	0.0102 V
TC	0 V	± 2 mV/°C	95	1.020 mV/°C	5.6 °C	0.0057 V
$TR_{°C}$	5.6 °C	± 1.1 °C	99	0.427°C	0	0 V

From equation (D-1), the nominal amplifier output is computed to be

$$Amp_{out} = LC_{out} \times G = 9.60 \text{ mV} \times 0.5 \text{ V/mV} = 4.80 \text{ V}.$$

The total uncertainty in the amplifier output voltage is computed by taking the root sum square of the component uncertainties.

$$u_{Amp_{out}} = \sqrt{\begin{aligned}&(0.0869 \text{ V})^2 + (0.0026 \text{ V})^2 + (0.0005 \text{ V})^2 + (0.0005 \text{ V})^2 \\ &+ (0.0006 \text{ V})^2 + (0.0102 \text{ V})^2 + (0.0057 \text{ V})^2\end{aligned}}$$

$$= \sqrt{0.0077 \text{ V}^2}$$

$$= 0.0877 \text{ V} = 87.7 \text{ mV}.$$

The degrees of freedom for the uncertainty in the amplifier output voltage is computed using the Welch-Satterthwaite formula, as shown in equation (D-7).

$$v_{Amp_{out}} = \frac{u_{Amp_{out}}^4}{\sum_{i=1}^{8} \dfrac{c_i^4 u_i^4}{v_i}} \tag{D-7}$$

where

$$\sum_{i=1}^{8} \frac{c_i^4 u_i^4}{v_i} = \frac{c_{LC_{out}}^4 u_{LC_{out}}^4}{v_{LC_{out}}} + \frac{c_{G_{Acc}}^4 u_{G_{Acc}}^4}{v_{G_{Acc}}} + \frac{c_{G_S}^4 u_{G_S}^4}{v_{G_S}} + \frac{c_{G_{NL}}^4 u_{G_{NL}}^4}{v_{G_{NL}}} + \frac{c_{G_{NS}}^4 u_{G_{NS}}^4}{v_{G_{NS}}}$$

$$+ \frac{c_{BST}^4 u_{BST}^4}{v_{BST}} + \frac{c_{TC}^4 u_{TC}^4}{v_{TC}} + \frac{c_{TR_{°C}}^4 u_{TR_{°C}}^4}{v_{TR_{°C}}}$$

The degrees of freedom for all of the error source uncertainties are assumed infinite. Therefore, the degrees of freedom for the uncertainty in the amplifier output is infinite.

D.3   Digital Multimeter Module (M₃)

The third system module is a model 8602A digital multimeter, manufactured by Fluke. This module converts the analog output signal from the amplifier module to a digital signal and displays it on a readout device. The basic transfer function for this module is expressed in equation (D-8).

$$DMM_{out} = Amp_{out} \qquad \text{(D-8)}$$

where

DMM_{out} = Digital multimeter output, V

Manufacturer's published specifications for the DC voltage function of the digital multimeter[71] are listed in Table D-5. In this module, key error sources include:

- DC voltmeter accuracy
- DC voltmeter digital resolution

Table D-5. DC Voltage Specifications for 8062A Multimeter

Specification	Value	Units
20 V Range Resolution	1	mV
20 V Range Accuracy	0.07% of Reading + 2 digits	mV

DC Voltage Accuracy.
The overall accuracy of the DC Voltage reading for a 20 V range is specified as ± (0.07% of reading + 2 digits). This specification is interpreted to be a 95% confidence interval for normally distributed errors.

Digital Resolution.
The digital resolution for the 20 V DC range is specified as 1 mV. Since this is a digital display, the resolution error is uniformly distributed. Therefore, the resolution error limits ± 0.5 mV are interpreted to be the minimum 100% containment limits.

The digital multimeter output equation must be modified before the associated error model can be developed. The appropriate module output equation given in equation (D-9) accounts for the relevant module parameters and error limits listed in Table D-6.

$$DMM_{out} = Amp_{out} + DMM_{Acc} + DMM_{res} \qquad \text{(D-9)}$$

Table D-6. Parameters used in Modified Multimeter Module Equation

Param. Name	Description	Nominal or Mean Value	Error Limits	Percent Confidence
Amp_{out}	DMM Input	4.80 V		
DMM_{Acc}	DC Voltmeter Accuracy	0 V	± (0.07% Read + 2 mV)	95
DMM_{res}	DC Voltmeter Digital Resolution	0 V	± 0.5 mV	100

The corresponding error model for the multimeter module is given in equation (D-10).

$$\varepsilon_{DMM_{out}} = c_{Amp_{out}} \varepsilon_{Amp_{out}} + c_{DMM_{Acc}} \varepsilon_{DMM_{Acc}} + c_{DMM_{res}} \varepsilon_{DMM_{res}} \qquad \text{(D-10)}$$

---

[71] Specifications from 8062A Instruction Manual downloaded from www.fluke.com

The partial derivative equations used to compute the sensitivity coefficients are listed below.

$$c_{Amp_{out}} = \frac{\partial DMM_{out}}{\partial Amp_{out}} = 1 \qquad c_{DMM_{Acc}} = \frac{\partial DMM_{out}}{\partial DMM_{Acc}} = 1 \qquad c_{DMM_{res}} = \frac{\partial DMM_{out}}{\partial DMM_{res}} = 1$$

The uncertainty model for the multimeter module output is developed by applying the variance operator to the error model.

$$
\begin{aligned}
u_{DMM_{out}} &= \sqrt{\mathrm{var}\left(\varepsilon_{DMM_{out}}\right)} \\
&= \sqrt{\mathrm{var}\left(c_{Amp_{out}}\varepsilon_{Amp_{out}} + c_{DMM_{Acc}}\varepsilon_{DMM_{Acc}} + c_{DMM_{res}}\varepsilon_{DMM_{res}}\right)} \qquad\text{(D-11)} \\
&= \sqrt{\mathrm{var}\left(\varepsilon_{Amp_{out}} + \varepsilon_{DMM_{Acc}} + \varepsilon_{DMM_{res}}\right)}
\end{aligned}
$$

There are no correlations between error sources, so the uncertainty model for the multimeter module output can be expressed as

$$u_{DMM_{out}} = \sqrt{u_{Amp_{out}}^2 + u_{DMM_{Acc}}^2 + u_{DMM_{res}}^2} \qquad\text{(D-12)}$$

The multimeter accuracy error follows a normal distribution, so the uncertainty estimated to be

$$
\begin{aligned}
u_{DMM_{Acc}} &= \frac{\left(4.8\,\mathrm{V}\times\dfrac{0.07}{100}\times\dfrac{1000\,\mathrm{mV}}{1\,\mathrm{V}} + 2\,\mathrm{mV}\right)}{\Phi^{-1}\left(\dfrac{1+0.95}{2}\right)} = \left(\frac{3.4\,\mathrm{mV} + 2\,\mathrm{mV}}{1.9600}\right) \\
&= \frac{5.4\,\mathrm{mV}}{1.9600} = 2.7\,\mathrm{mV}.
\end{aligned}
$$

The multimeter resolution error follows a uniform distribution, so the uncertainty is estimated to be

$$
\begin{aligned}
u_{DMM_{res}} &= \frac{0.5\,\mathrm{mV}}{\sqrt{3}} = \frac{0.5\,\mathrm{mV}}{1.732} \\
&= 0.3\,\mathrm{mV}.
\end{aligned}
$$

The estimated uncertainties for each parameter are listed in Table D-7.

Table D-7. Estimated Uncertainties for Digital Multimeter Module Parameters

Param. Name	Nominal or Mean Value	± Error Limits	Percent Conf.	Standard Uncertainty	Sensitivity Coefficient	Component Uncertainty
Amp_{out}	4.800 V			87.7 mV	1	87.7 mV
DMM_{Acc}	0 V	± 5.4 mV	95	2.7 mV	1	2.7 mV
DMM_{res}	0 V	± 0.5 mV	100	0.3 mV	1	0.3 mV

The output from the multimeter module is 4.800 V and the total uncertainty in this value is computed by taking the root sum square of the component uncertainties.

$$u_{DMM_{out}} = \sqrt{(87.7 \text{ mV})^2 + (2.7 \text{ mV})^2 + (0.3 \text{ mV})^2}$$
$$= \sqrt{7699 \text{ mV}^2}$$
$$= 87.74 \text{ mV}.$$

The degrees of is computed using equation (D-13).

$$v_{DMM_{out}} = \frac{u_{DMM_{out}}^4}{\dfrac{c_{Amp_{out}}^4 u_{Amp_{out}}^4}{v_{Amp_{out}}} + \dfrac{c_{DMM_{Acc}}^4 u_{DMM_{Acc}}^4}{v_{DMM_{Acc}}} + \dfrac{c_{DMM_{res}}^4 u_{DMM_{res}}^4}{v_{DMM_{res}}}} \tag{D-13}$$

The degrees of freedom for error source uncertainties were assumed to be infinite, so, the degrees of freedom for the estimated uncertainty in the multimeter output is computed to be infinite.

D.4   Data Processor Module (M4)

The last system module is the data processor module. The data processor converts the multimeter voltage output to a load value using a linear regression equation obtained from calibration data. The basic transfer function for this module is expressed in equation (D-14).

$$DP_{out} = c_1 \times DMM_{out} + c_2 \tag{D-14}$$

where

c_1 = regression line slope
c_2 = regression line intercept

Errors associated with data processing result from computation round-off or truncation and from residual differences between an observed value during calibration and the value estimated from the regression equation.[72] For this module, regression error must be considered. The appropriate module output equation is expressed as

$$DP_{out} = c_1 \times DMM_{out} + c_2 + reg \tag{D-15}$$

where reg is the standard error of forecast. In regression analysis, the standard error of forecast is a function of the standard error of estimate. Both quantities are discussed below.

Standard Error of Estimate
Standard error of estimate is a measure of the difference between actual values and values estimated from a regression equation.[73] A regression analysis that has a small standard error has

---

[72] Error sources resulting from data reduction and analysis are often overlooked in the assessment of measurement uncertainty.

[73] Hanke, J. et al.: *Statistical Decision Models for Management*, Allyn and Bacon, Inc. 1984.

data points that are very close to the regression line. Conversely, a large standard error results when data points are widely dispersed around the regression line. The standard error of estimate is computed using equation (D-16).

$$S_{y,x} = \sqrt{\frac{\sum (y - \hat{y})^2}{n-2}}$$

(D-16)

where $\hat{y}$ is the predicted value, y is the observed or measured value obtained during calibration, and n is the number of measured data points used to establish the regression equation. For the purposes of this example, the standard error of estimate is assumed to be equal to 0.02 lb$_f$.

Standard Error of Forecast

As previously stated, the standard error of estimate is a measurement of the typical vertical distance of the sample data points from the regression line. However, we must also consider the fact that the regression line was generated from a finite sample of data. If another sample of data were collected, then a different regression line would result. Therefore, we must also consider the dispersion of various regression lines that would be generated from multiple sample sets around the true population regression line.

The standard error of the forecast accounts for the dispersion of the regression lines and is computed using equation (D-17).

$$s_f = s_{y,x} \sqrt{1 + \frac{1}{n} + \frac{(x - \bar{x})^2}{\sum (x - \bar{x})^2}}$$

(D-17)

where $\bar{x}$ is the average of all values of x over the regression fit range. As seen from equation (D-17), a standard error of forecast is computed for each value of x. For a 3 lb$_f$ load used in this analysis, s_f has a value of 0.022 lb$_f$.

The error model for the data processor module is given in equation (D-18).

$$\varepsilon_{DP_{out}} = c_{DMM_{out}} \varepsilon_{DMM_{out}} + c_{reg} \varepsilon_{reg}$$

(D-18)

The partial derivative equations used to compute the sensitivity coefficients are listed below.

$$c_{DMM_{out}} = \frac{\partial DP_{out}}{\partial DMM_{out}} = c_1 \qquad c_{reg} = \frac{\partial DP_{out}}{\partial reg} = 1$$

The uncertainty model for the data processor module output is developed by applying the variance operator to the error model.

$$u_{DP_{out}} = \sqrt{\text{var}\left(\varepsilon_{DP_{out}}\right)} = \sqrt{\text{var}\left(c_{DMM_{out}} \varepsilon_{DMM_{out}} + c_{reg} \varepsilon_{reg}\right)}$$

(D-19)

There are no correlations between error sources, so the uncertainty model for the data processor

151

module output can be expressed as

$$u_{DP_{out}} = \sqrt{c_{DMM_{out}}^2 u_{DMM_{out}}^2 + c_{reg}^2 u_{reg}^2} \tag{D-20}$$

The uncertainty in the data processor output is computed from the uncertainty estimates and sensitivity coefficients for each parameter listed in Table D-8.

Table D-8. Estimated Uncertainties for Data Processor Module Parameters

Param. Name	Nominal or Mean Value	± Error Limits	Percent Conf.	Standard Uncertainty	Sensitivity Coefficient	Component Uncertainty
DMM_{out}	4.800 V			87.74 mV	0.621 lb$_f$/V	0.055 lb$_f$
c_1	0.621 lb$_f$/V					
c_2	0.015 lb$_f$					
reg	0 lb$_f$			0.022 lb$_f$	1	0.022 lb$_f$

The output from the data processor module is

$$DP_{out} = 0.621 \text{ lb}_f/\text{V} \times 4.80 \text{ V} + 0.015 \text{ lb}_f = 2.996 \text{ lb}_f$$

and the total uncertainty in this value is computed by taking the root sum square of the component uncertainties.

$$u_{DP_{out}} = \sqrt{\left(0.055 \text{ lb}_f\right)^2 + \left(0.022 \text{ lb}_f\right)^2} = \sqrt{0.00351 \text{ lb}_f^2} = 0.059 \text{ lb}_f.$$

The degrees of is computed using equation (D-21).

$$\nu_{DP_{out}} = \frac{u_{DP_{out}}^4}{\dfrac{c_{DMM_{out}}^4 u_{DMM_{out}}^4}{\nu_{DMM_{out}}} + \dfrac{c_{reg}^4 u_{reg}^4}{\nu_{reg}}} \tag{D-21}$$

The degrees of freedom for multimeter output uncertainty is infinite and the degrees of freedom for the regression error is 8, so the total degrees of freedom for the uncertainty in the data processor output is computed to be

$$\nu_{DP_{out}} = \frac{\left(0.059 \text{ lb}_f\right)^4}{\dfrac{\left(0.055 \text{ lb}_f\right)^4}{\infty} + \dfrac{\left(0.022 \text{ lb}_f\right)^4}{8}} = \frac{1.212 \times 10^{-5} \text{ lb}_f^4}{\dfrac{0.884 \times 10^{-5} \text{ lb}_f^4}{\infty} + \dfrac{0.023 \times 10^{-5} \text{ lb}_f^4}{8}}$$

$$= \frac{1.212}{0.023} \times 8 = 422 \cong \infty$$

D.5   System Output Uncertainty

The analysis results for the load measurement system are summarized in Table D-9. The system output uncertainty is equal to the output uncertainty for the final module. The associated degrees of freedom for the system output uncertainty is also equal to the degrees of freedom for the final module output uncertainty.

Table D-9. Summary of Results for Load Measurement System Analysis

Module Name	Module Input	Module Output	Standard Uncertainty	Degrees of Freedom
Load Cell	3 lb$_f$	9.600 mV	0.174 mV	∞
Amplifier	9.600 mV	4.800 V	87.7 mV	∞
Digital Multimeter	4.800 V	4.800 V	87.8 mV	∞
Data Processor Module	4.800 V	2.996 lb$_f$	0.059 lb$_f$	∞

For a 3 lb$_f$ input load, the system output and uncertainty are 2.996 lb$_f$ and 0.059 lb$_f$, respectively. The confidence limits have not yet been computed, but it is apparent that the ± 5 lb$_f$ requirement cannot be meet with this measurement system design. The amplifier module increases the uncertainty substantially. However, it is useful to take a closer look to determine how the error source uncertainties for each module contribute to the overall system output uncertainty. This is accomplished by viewing the Pareto chart for each module, shown in Figures D-2 through D-5.

The Pareto chart for the load cell shows that the excitation voltage and zero offset errors are the largest contributors to the module output uncertainty. Because the uncertainty in the load cell output is multiplied by the amplifier gain, it is the largest contributor to the overall uncertainty in the amplifier output, as shown in Figure D-2. Similarly, the uncertainty in the amplifier output is the largest contributor to the overall uncertainty in the digital multimeter output.

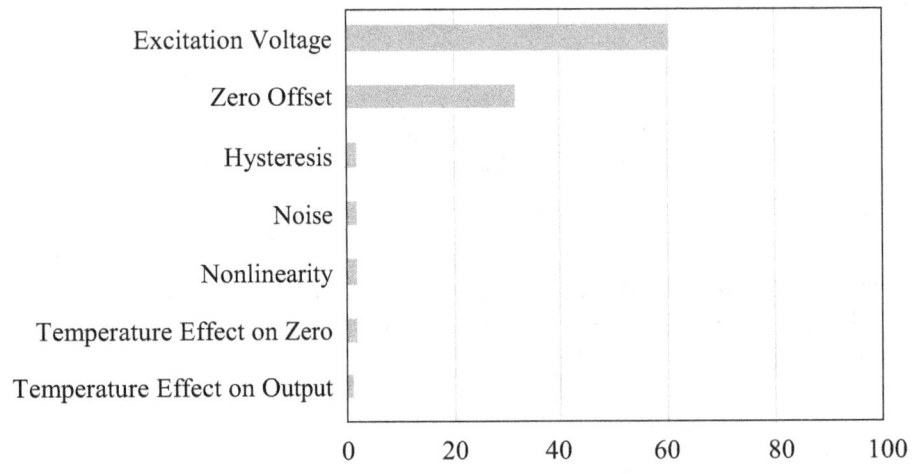

Percent Contribution to Uncertainty in Load Cell Output

Figure D-2 Pareto Chart for Load Cell Module

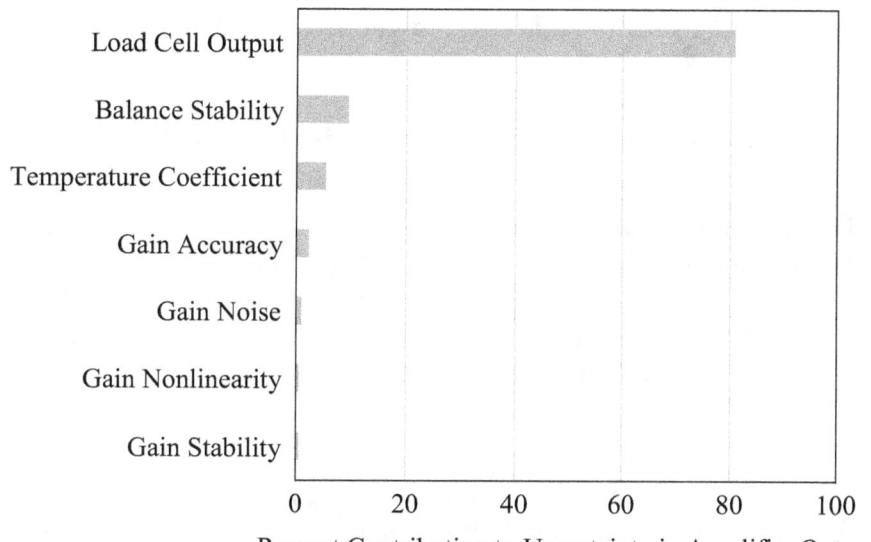

Percent Contribution to Uncertainty in Amplifier Output

Figure D-3 Pareto Chart for Amplifier Module

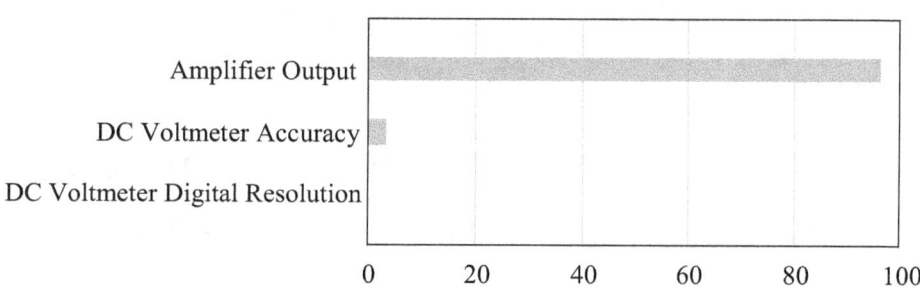

Percent Contribution to Uncertainty in Multimeter Output

Figure D-4 Pareto Chart for Digital Multimeter Module

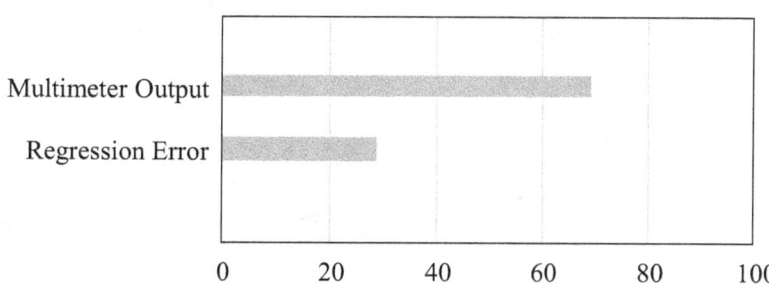

Percent Contribution to Uncertainty in Data Processor Output

Figure D-5 Pareto Chart for Data Processor Module

The uncertainty in the multimeter output is the largest contributor to the uncertainty in the data processor output. The Pareto charts illustrate how the uncertainty in the load cell output propagates through the subsequent modules and is the largest contributor to the overall system output uncertainty.

A more accurate voltage source could significantly reduce the uncertainty in the load cell output. This in-turn, would reduce the uncertainty in the system output. For example, a precision power

154

supply with specifications[74] listed in Table D-10 might be considered as the excitation voltage source for the load cell.

Table D-10. Specifications for 1785B DC Power Supply

Specification	Value	Units
Output Rating (0 to 40 °C)	0 to 18	V
Load Regulation	± (0.01% of output + 3)	mV
Line Regulation	± (0.1% of output + 3)	mV
Ripple (20Hz to 20 MHz)	< 3	mV
Digital Display Resolution	10	mV

Load Regulation

Load regulation is a measure of the ability of the power supply to maintain a constant output voltage output when a device is connected to it that has a load resistance, R_L. In constant voltage mode, the load regulation specification defines how close the series resistance, R_S, of the output is to 0 ohms (the series resistance of an ideal voltage source, V_S).

$$V_{Ex} = V_s \frac{R_L}{R_L + R_s}$$

For an 8 VDC output, the manufacturer specification of ± (0.01% of output + 3 mV) is

$$\pm (0.01/100 \times 8000 \text{ mV} + 3 \text{ mV}) = \pm (0.80 \text{ mV} + 3 \text{ mV}) = \pm 3.8 \text{ mV}$$

These limits are interpreted to be a 95% confidence interval for normally distributed errors.

Line Regulation

Line regulation is a measure of the ability of the power supply to maintain its output voltage given changes in the input line voltage. For a 8 VDC output, the manufacturer specification of ± (0.1% of output + 3 mV) is

$$\pm (0.1/100 \times 8000 \text{ mV} + 3 \text{ mV}) = \pm (8 \text{ mV} + 3 \text{ mV}) = \pm 11 \text{ mV}$$

These limits are interpreted to be a 95% confidence interval for normally distributed errors.

Ripple

Ripple and noise is random error intrinsic to the power supply and is usually specify in absolute units of volts RMS or volts peak-to-peak. The manufacturer specification of 3 mV peak-to-peak is estimated to be ± 1.5 mV limits that are equivalent to a 99% confidence interval for normally distributed errors.

Digital Resolution

The resolution for the setting the output power is 10 mV. Since this is a digital display, the resolution error is uniformly distributed. Therefore, the resolution error limits of ± 5 mV are interpreted to be the minimum 100% containment limits.

---

[74] Specifications obtained from www.bkprecision.com

The error model for the DC voltage output from the model 1785B power supply is given in equation (D-22).

$$\varepsilon_{V_{Ex}} = \varepsilon_{LD} + \varepsilon_{LN} + \varepsilon_{RP} + \varepsilon_{res} \qquad \text{(D-22)}$$

The parameters used in the error model equation are listed in Table D-11.

Table D-11. Parameters used in Power Supply Error Model

Parameter Name	Description	Nominal or Mean Value	Error Limits	Percent Confidence
ε_{LD}	Load Regulation Error	0 V	± 3.8 mV	95
ε_{LN}	Line Regulation Error	0 V	± 11 mV	95
ε_{RP}	Ripple and Noise Error	0 V	± 1.5 mV	99
ε_{res}	Digital Resolution Error	0 V	± 5 mV	100

The uncertainty model for the DC voltage output is developed by applying the variance operator to the error model.

$$u_{V_{Ex}} = \sqrt{\mathrm{var}\left(\varepsilon_{V_{Ex}}\right)} = \sqrt{\mathrm{var}\left(\varepsilon_{LD} + \varepsilon_{LN} + \varepsilon_{RP} + \varepsilon_{res}\right)} \qquad \text{(D-23)}$$

There are no correlations between error sources, so the uncertainty model for the voltage output can be expressed as

$$u_{V_{Ex}} = \sqrt{u_{LD}^2 + u_{LN}^2 + u_{RP}^2 + u_{res}^2} \qquad \text{(D-24)}$$

The load regulation, line regulation and ripple error sources all follow a normal distribution, so the corresponding uncertainties are estimated to be

$$u_{LD} = \frac{(3.8\,\text{mV})}{\Phi^{-1}\left(\dfrac{1+0.95}{2}\right)} = \left(\frac{3.8\,\text{mV}}{1.9600}\right) = 1.94\,\text{mV}.$$

$$u_{LN} = \frac{(11\,\text{mV})}{\Phi^{-1}\left(\dfrac{1+0.95}{2}\right)} = \left(\frac{11\,\text{mV}}{1.9600}\right) = 5.61\,\text{mV}.$$

$$u_{RP} = \frac{(1.5\,\text{mV})}{\Phi^{-1}\left(\dfrac{1+0.99}{2}\right)} = \left(\frac{1.5\,\text{mV}}{2.576}\right) = 0.58\,\text{mV}.$$

The resolution error follows a uniform distribution, so the uncertainty is estimated to be

$$u_{res} = \frac{5\text{ mV}}{\sqrt{3}} = \frac{5\text{ mV}}{1.732} = 2.89 \text{ mV}.$$

The estimated uncertainties for each parameter are summarized in Table D-12.

Table D-12. Estimated Uncertainties for Power Supply Parameters

Param. Name	Nominal or Mean Value	± Error Limits	Percent Conf.	Standard Uncertainty
ε_{LD}	0 V	± 3.8 mV	95	1.94 mV
ε_{LN}	0 V	± 11 mV	95	5.61 mV
ε_{RP}	0 V	± 1.5 mV	99	0.58 mV
ε_{res}	0 V	± 5 mV	100	2.89 mV

The total uncertainty in the 8 DC voltage output is computed by taking the root sum square of the standard uncertainties.

$$u_{V_{Ex}} = \sqrt{(1.94 \text{ mV})^2 + (5.61 \text{ mV})^2 + (0.58 \text{ mV})^2 + (2.89 \text{ mV})^2}$$

$$= \sqrt{43.93 \text{ mV}^2} = 6.63 \text{ mV}.$$

With the more accurate excitation voltage source, the uncertainty in the load cell output is computed by taking the root sum square of the component uncertainties listed in Table D-13.

Table D-13. Estimated Uncertainties for Load Cell Parameters

Param. Name	Nominal or Stated Value	± Error Limits	Conf. Level	Standard Uncertainty	Sensitivity Coefficient	Component Uncertainty
NL	0 mV/V	± 0.001 mV/V	95	0.0005 mV/V	8 V	0.0041 mV
Hys	0 mV/V	± 0.001 mV/V	95	0.0005 mV/V	8 V	0.0041 mV
NS	0 mV/V	± 0.001 mV/V	95	0.0005 mV/V	8 V	0.0041 mV
ZO	0 mV/V	± 0.02 mV/V	95	0.0102 mV/V	8 V	0.0816 mV
TE_{out}	0 lb$_f$/°F	± 1.5 × 10^{-4} lb$_f$/°F	95	7.65 × 10^{-5} lb$_f$/°F	32 °F × mV/lb$_f$	0.0024 mV
TE_{zero}	0 mV/°F	± 0.0001 mV/V/°F	95	5 × 10^{-5} mV/V/°F	80 °F × V	0.0041 mV
V_{Ex}	8 V			0.00663 V	1.2 mV/V	0.0080 mV

$$u_{LC_{out}} = \sqrt{\begin{array}{l}(0.0041 \text{ mV})^2 + (0.0041 \text{ mV})^2 + (0.0041 \text{ mV})^2 + (0.0816 \text{ mV})^2 \\ + (0.0024 \text{ mV})^2 + (0.0041 \text{ mV})^2 + (0.0080 \text{ mV})^2\end{array}}$$

$$= \sqrt{0.0068 \text{ mV}^2} = 0.0824 \text{ mV}.$$

By using a more accurate excitation voltage source, the total uncertainty in the load cell output is reduced by over 50%. Propagating the new load cell output uncertainty through to the amplifier module, the uncertainty in the amplifier output is computed to be

$$u_{Amp_{out}} = \sqrt{\begin{array}{l}(0.5\ \text{V/mV} \times 0.0824\ \text{mV})^2 + (0.0026\ \text{V})^2 + (0.0005\ \text{V})^2 + (0.0005\ \text{V})^2 \\ + (0.0006\ \text{V})^2 + (0.0102\ \text{V})^2 + (0.0057\ \text{V})^2\end{array}}$$

$$= \sqrt{0.00184\ \text{V}^2}$$

$$= 0.0429\ \text{V} = 42.9\ \text{mV}.$$

Similarly, the uncertainties in the multimeter output and data processor output are computed to be

$$u_{DMM_{out}} = \sqrt{(42.9\ \text{mV})^2 + (2.7\ \text{mV})^2 + (0.3\ \text{mV})^2}$$

$$= \sqrt{1847.8\ \text{mV}^2} = 43\ \text{mV}$$

and

$$u_{DP_{out}} = \sqrt{(0.621\ \text{lb}_f/\text{V} \times 43\ \text{mV} \times 0.001\ \text{V/mV})^2 + (0.022\ \text{lb}_f)^2}$$

$$\sqrt{(0.027\ \text{lb}_f)^2 + (0.022\ \text{lb}_f)^2}$$

$$= \sqrt{0.00121\ \text{lb}_f^2} = 0.0348\ \text{lb}_f.$$

For a 3 lb$_f$ input load, the system output and uncertainty are now 2.996 lb$_f$ and 0.0348 lb$_f$, respectively. This system analysis can be duplicated for other applied loads. The results, listed in Table D-14, show that the uncertainty in the measurement system output is basically constant over the specified range of input loads.

Table D-14. Uncertainty in System Output versus Applied Load

Applied Load	Output Load	Total Uncertainty
1 lb$_f$	1.01 lb$_f$	0.0347 lb$_f$
2 lb$_f$	2.00 lb$_f$	0.0347 lb$_f$
3 lb$_f$	3.00 lb$_f$	0.0348 lb$_f$
4 lb$_f$	3.99 lb$_f$	0.0349 lb$_f$
5 lb$_f$	4.98 lb$_f$	0.0351 lb$_f$

D.6   Confidence Limits

The system output uncertainty and degrees of freedom can be used to compute confidence limits that are expected to contain the system output with some specified confidence level or probability, p. The confidence limits are expressed as

$$\pm t_{\alpha/2,\nu} u_{DP_{out}} \tag{D-25}$$

where the multiplier, $t_{\alpha/2\nu}$, is the t-statistic and $\alpha = 1 - p$.

In this example, the 99% confidence limits (i.e., $p = 0.99$) for the load measurement system are required. The corresponding t-statistic is $t_{0.005,\infty} = 2.576$ and the confidence limits are computed to be

$$\pm\, 2.576 \times 0.035 \text{ lb}_f = \pm\, 0.09 \text{ lb}_f.$$

These limits do not meet the required 99% confidence limits of $\pm\, 0.05$ lb$_f$. For this measurement system, the $\pm\, 0.05$ lb$_f$ limits correspond to an 85% confidence level, where $t_{0.075,\infty} = 1.44$.

$$\pm\, 1.44 \times 0.035 \text{ lb}_f = \pm\, 0.05 \text{ lb}_f$$

Given the results of this analysis, the options are to either modify the measurement requirements (i.e., reduce the confidence level or increase the confidence limits) or select different components to further reduce the system output uncertainty.

REFERENCES

1. Allegro MicroSystems, Inc.: "General Information: A Complete Guide to Data Sheets," Publication 26000A, 1998.

2. ANSI/ASQC Z1.4-2003 *Sampling Procedures and Tables for Inspection by Attributes*.

3. ANSI/ASQC Z1.9-2003 *Sampling Procedures and Tables for Inspection by Variables for Percent Nonconforming*.

4. ANSI/ISA 51.1-1979 (R1993), *Process Instrumentation Terminology*, The Instrumentation, Systems and Automation Society, April 1979.

5. ANSI/NCSL Z540.1 (R2002), *Calibration Laboratories and Measuring and Test Equipment – General Requirements*, July 1994.

6. ANSI/NCSLI Z540.3: 2006 *Requirements for the Calibration of Measuring and Test Equipment*.

7. ASME B89.7.4.1-2005 *Measurement Uncertainty and Conformance Testing: Risk Analysis*, American Society of Mechanical Engineers, February 3, 2006.

8. ASTM E2587 - 07 *Standard Practice for Use of Control Charts in Statistical Process Control*, 2007.

9. ASTM E2281 - 08 *Standard Practice for Process and Measurement Capability Indices*, 2008.

10. Automotive Industry Action Group, *Measurement Systems Analysis Reference Manual*, 3rd Edition, 2003.

11. Castrup, H.: "Selecting and Applying Error Distributions in Uncertainty Analysis," presented at the Measurement Science Conference, Anaheim, CA, 2004.

12. Castrup, H.: "Risk Analysis Methods for Complying with NCSLI Z540.3," Proceedings of the NCSLI International Workshop and Symposium, St. Paul, MN August 2007.

13. Crisp, P. B.: "DMM Terminology," Cal Lab Magazine, July-August, 1997.

14. Deaver, D.: "Having Confidence in Specifications," proceeding of NCSLI Workshop and Symposium, Salt Lake City, UT, July 2004.

15. Fluke Corporation: *Calibration: Philosophy in Practice*, 2nd Edition, 1994.

16. Fluke Corporation: *Precision Measurement Solutions*, 2004.

17. Goldberg, B. E., et al.: *NASA System Engineering "Toolbox" for Design-Oriented Engineers*, NASA Reference Publication 1358, December 1994.

18. Hanke, J. et al.: *Statistical Decision Models for Management*, Allyn and Bacon, Inc. 1984.

19. ISA-37.1-1975 (R1982): *Electrical Transducer Nomenclature and Terminology*, The Instrumentation, Systems and Automation Society, Reaffirmed December 14, 1982.

20. ISA-RP37.3-1982-(R1995): *Specifications and Tests for Strain Gauge Pressure Transducers*, The Instrumentation, Systems and Automation Society.

21. ISO 2859-0:1995 *Sampling procedures for inspection by attributes - Part 0: Introduction to the ISO 2859 attribute sampling system.*

22. ISO 2859-1:1999 *Sampling procedures for inspection by attributes - Part 1: Sampling plans indexed by acceptable quality level (AQL) for lot-by-lot inspection.*

23. J. S. Wilson (Editor-in-Chief): *Sensor Technology Handbook*, Elsevier, 2005.

24. Keithley Instruments: *Low Level Measurements Handbook*, 6th Edition.

25. Kleman, K. S.: "The Origin of Specification," American Chemical Society, 2001 (www.pubs.acs.org).

26. McGee, T. D.: *Principles and Methods of Temperature Measurement*, John Wiley & Sons, 1988.

27. MIL-STD-105E: *Sampling Procedures and Tables for Inspection by Attributes*, May 10, 1989.

28. MIL-STD-202G, *Department of Defense Test Method Standard Electronic and Electrical Component Parts*, February 8, 2002.

29. MIL-STD-781: *Reliability Testing for Engineering Development, Qualification and Production*, October 1986.

30. MIL-STD-810F: *Department of Defense Test Method Standard for Environmental Engineering Considerations and Laboratory Tests*, January 2000.

31. MIL-STD-1540D: *Product Verification Requirements for Launch, Upper Stage, and Space Vehicles*, Department of Defense Standard Practice, January 15, 1999.

32. MIL-STD-1629A: *Procedures for Performing a Failure Mode, Effects and Criticality Analysis*, Department of Defense Military Standard, November 24, 1980.

33. MIL-STD 45662A, *Calibration Systems Requirements*, U.S. Dept. of Defense, 1 August 1988.

34. NASA-HNBK-8739.19 NASA *Measurement Quality Assurance Handbook*

35. NASA-HNBK-8739.19-3 *Measurement Quality Assurance Handbook* Annex 3 – *Measurement Uncertainty Analysis Principles and Methods.*

36. NASA-HNBK-8739.19-4 *Measurement Quality Assurance Handbook* Annex 4 – *Estimation and Evaluation of Measurement Decision Risk.*

37. NASA-HNBK-8739.19-5 *Measurement Quality Handbook* Annex 5 – *Establishment and Adjustment of Calibration Intervals.*

38. NASA SP-T-0023 Revision C, *Space Shuttle Specification Environmental Acceptance Testing*, May 17, 2001.

39. National Instruments: "Data Acquisition Specifications – a Glossary," Application Note 092, February 1997.

40. National Instruments: "Demystifying Instrument Specifications – How to Make Sense Out of the Jargon," Application Note 155, December 2000.

41. NCSLI RP-3 *Calibration Procedures*, October 2007.

42. NIST Monograph 180, *Gage Block Handbook*, U.S. National Institute of Standards and Technology, 2004.

43. NIST/SEMATEC: *e-Handbook of Statistical Methods*, 7/18/2006 http://www.itl.nist.gov/div898/handbook/.

44. *Numerical Recipes in FORTRAN: The Art of Scientific Computing*, 2nd Edition, Cambridge University Press, 1992.

45. SMA LCS 04-99: *Standard Load Cell Specifications*, Scale Manufacturers Association, Provisional 1st Edition, April 24, 1999.

46. Spitzer, D. W.: *Industrial Flow Measurement*, ISA, 1990.

47. Taylor, J. L.: *Computer-Based Data Acquisition Systems Design Techniques*, Instrument Society of America, 1986.

48. TM 5-698-4: *Failure Modes, Effects and Criticality Analysis (FMECA) for Command, Control, Communications, Computer, Intelligence, Surveillance, and Reconnaissance (C4ISR) Facilities*, Headquarters, Department of the Army, September 29, 2006.

49. Webster, J. G. (Ed.): *Measurement, Instrumentation and Sensors Handbook*, Chapman & Hall/CRCnetBase, 1999.

www.ingramcontent.com/pod-product-compliance
Lightning Source LLC
Chambersburg PA
CBHW081723220526
45468CB00008B/1954